THE AI-HUMAN PARADOX

WHAT ARTIFICIAL INTELLIGENCE REVEALS ABOUT CONSCIOUSNESS AND THE FUTURE OF BEING HUMAN

By

Bruce D. Randall

Book One in the Paradox Series

ALSO BY BRUCE D. RANDALL

The AI Human Paradox

What Artificial Intelligence Reveals

About Consciousness and the Future of Being Human

The Quantum Human Paradox

When Quantum Computing Changes Everything:

The Human Questions Machines Can't Answer

The QAI Human Paradox

Where Artificial Intelligence Meets Quantum Reality:

A New Map for Consciousness and the Future of Being Human

The End of Smart

The Human Edge in the Age of AI:

Why Judgment, Presence, and Wisdom Now Outperform Raw Intelligence

Still Present

What I Found When Certainty Fell Away

Enoch and the Machine

The Book of Enoch, the Nephilim,

and the Ancient Pattern Behind the Age of Artificial Intelligence

ISBNs:
eBook: 978-1-960346-93-3
Paperback: 978-1-960346-94-0
Hardback: 978-1-960346-95-7

AUTHORSUNITE

Published by Authors Unite

My Love extends *to Carine for supporting me throughout the late nights and weekends during this journey.*

My gratitude extends *to Sandi Andersen for her support during the writing of this book.*

My gratitude and thankfulness extend *to Josh Reid for his contribution to a chapter and his consultation on this idea.*

Grateful *for the friends and family that have stood by me during this first book of my trilogy with two more books to go.*

Also thankful *to Leap Academy for helping me through the 2020 life changes while*

I found my new calling with podcasts, writing books, and the new Executive position that I've been blessed to secure.

TABLE OF CONTENTS

Prologue:

WHY THIS BOOK, WHY NOW?

Over millennia, humankind has advanced through its phases of development—from cave dwellers to hunter gatherers, then through the discovery of tools such as the wheel and the inclined plane, and advancing through the iron and bronze ages, and beyond. Yet, rarely, if ever, has any of the technological advancements discovered by individuals and adopted at scale posed a threat to civilization itself. With the rare exception of three advancements, no other single advancement has posed a perceived threat on such a grand scale.

The discoveries of moveable type and the printing press were the first two to radically effect humankind. Yet these only threatened the power structure of a church that had morphed from championing spiritual advancement to becoming a political power that influenced the world. Yet, still they were a threat to the world of that era. The third of these has been the much more recent discovery and development of the atomic bomb, which truly positioned itself as an existential threat to the human race. Other than these three, few if any other technological advancements over the past several centuries have posed a threat to our understanding of who we are as, and what it means to be, human beings.

However, for those of us alive now, we see the world around us shifting faster than most of us can comprehend. Artificial intelligence, once the subject of speculative fiction, is now embedded into our daily lives—recommending what we watch, optimizing how businesses run, and even helping us design medicines that save lives. However, beneath the convenience and wonder that AI provides lies a deeper transformation: the convergence of AI, brain technology, and human consciousness. This is not merely a technological *evolution*; it is a *revolution* that challenges the core of what it means to be human.

We are at the edge of a paradox. Machines that were once tools are suddenly becoming collaborators, and in some cases, even contenders. As AI accelerates toward artificial general intelligence (AGI), we must begin to collectively ask ourselves questions that our ancestors have never faced: *What happens when our creations begin to think, reflect, and perhaps even feel? What do we lose, and what do we gain, when code begins to mirror consciousness?*

For me, the urgency of writing this book came about not in an abstract moment, but in real encounters. Watching AI compose music, argue philosophy, or detect cancers invisible to trained doctors no longer feels like science fiction. The future is not coming; it has already arrived—seeping into boardrooms, classrooms, hospitals, and homes. And yet, most of us are unprepared to navigate the moral, philosophical, psychological, and spiritual questions that this revolutionary new era boldly raises.

This book is neither a technical manual for engineers, nor is it an alarmist manifesto predicting the end of humanity. Instead, it is an invitation; to clearly view the world that is unfolding, *understand* its implications, and *engage in the greatest debate of our time*: how to shape intelligence that is no longer bound by our cranial protoplasm.

By the end of this journey, my hope is that you will not only understand the technologies shaping our century but also see yourself reflected in them. In the mirror of machine intelligence, we glimpse not only the future of AI—but the future of humanity itself.

This book has been written to serve as both a guide and a challenge. To the forces that are determined to transform our world—AI, neuroscience, and the redefinition of what it means to be intelligent—it is a guide. To those who think beyond the immediate utility of tools and dare to engage with the moral, social, and spiritual stakes of what we are building—it is also a challenge.

The paradox is already here. Our creations are not only amplifying our capabilities, but also reshaping our identities. The question is no longer whether or not AI will affect our lives. The question now is: *Are we ready to confront the world that AI is creating—and the individuals that it will require us to become?*

PART I:

WAKING THE MACHINE

Chapter One

FROM TOOLS TO MINDS

The dream of building machines that think like humans stretches back over centuries, long before computers, philosophers and inventors imagined mechanical minds. In the 17th century, Gottfried Wilhelm Leibniz — the co-inventor of calculus — envisioned a "universal language" that could encode all human reasoning. If such a symbolic system existed, he argued, disputes between parties could be settled by calculation: "let us calculate!".

Developments in the 20th Century provided the *intellectual scaffolding* for what we now know as artificial intelligence (AI). In 1936, British mathematician Alan Turing proposed that a "universal machine" could, in principle, simulate any form of computation. His work became the blueprint for digital computers. Turing later posed a provocative question: *Can machines think?* In his 1950 paper *Computing Machinery and Intelligence*, he introduced the "Imitation Game" — now called the Turing Test — as a pragmatic means of evaluating machine intelligence.

This was AI's intellectual spark, yet turning theory into practice would require the rise of digital computing, which materialized after World War II. Early pioneers the likes of John von Neumann, Norbert Wiener, and Claude Shannon laid the groundwork for this leap by establishing the necessary mathematical and engineering principles. As a result, in the mid 1950s the first programs capable of reasoning tasks appeared when Allen Newell and Herbert Simon's collaboration at RAND Corporation resulted in *Logic Theorist*, a revolutionary program that could prove elementary theorems in symbolic logic. During that same year — at Dartmouth College — John McCarthy, Marvin Minsky, Nathaniel Rochester, and Claude Shannon convened a historic workshop in which the term *artificial intelligence* was formally coined.

Due to this progress in such a brief period of time, optimism was high. McCarthy and colleagues boldly predicted that "every aspect of learning or any other feature of intelligence can in principle be so precisely described that a machine can be made to simulate it." During those heady early days, the belief was that human-level AI might arrive within a generation.

AI: ITS BEGINNINGS AND EVOLUTION

The 1950s and 1960s saw rapid experimentation. Programs could *play checkers* at competitive levels, solve algebraic equations, and even *translate* between Russian and English in limited domains. Due to the promise that intelligent machines could be harnessed for both national defense and economic progress, governments poured untold amounts of funding into AI research. This was particularly true in the U.S. Department of Defense through DARPA.

Yet reality soon undercut the hype. Early "symbolic AI" relied on hand-coded rules. During the decade of the 1950s, hand-coded rules (which were also called Good Old-Fashioned AI, and relied heavily on hand-coded rules) were the dominant approach to AI at the time. While systems performed impressively in constrained settings, they broke down in the complexity of 'real-world' applications. Translation systems stumbled over nuance and ambiguity. Programs that could solve 'toy problems' in logic, faltered in everyday reasoning.

By the 1970s disappointment had set in. As a result, funding dried up in what became known as the first AI winter. In spite of this occurrence, researchers kept working while the dream of imminent human-level AI faded from public view.

In the 1980s AI roared back with the rise of expert systems; programs that encoded specialized human knowledge into rule-based frameworks. For instance, MYCIN, developed at Stanford, could recommend medical treatments for infections based on symptoms and test results. Businesses also embraced such expert systems for tasks like configuring computer orders (e.g., DEC's XCON system).

Once again, optimism surged; venture capital flowed, startups multiplied, and corporate AI divisions expanded. By the mid-1980s, such 'expert systems' were marketed as being the future of business and industry.

In spite of all of the optimism and hype, there were limitations that persisted. For one, expert systems required enormous manual effort to encode knowledge. Secondly, they were brittle; if a situation fell outside their programmed rules, they collapsed. By the late 1980s AI disillusionment had returned. The market for expert systems contracted, and another AI winter followed.

With the appearance of the late 1990s and early 2000s came a crucial shift: instead of hand-coding rules, programmers let machines 'learn' patterns from data. With the rise of what was now being called machine learning (ML), AI was transformed into a statistical discipline. Algorithms the likes of decision trees, support vector machines, and Bayesian networks enabled systems to classify, predict, and cluster data at scale.

As computing power increased and the internet produced massive datasets, this statistical turn gained momentum. Instead of teaching a computer "what is a cat" with rules, engineers now 'trained' models on millions of labeled images. As a result, the model discovered its own statistical features.

Then in 2012 the breakthrough moment arrived with deep learning — multilayered, artificial 'neural networks' that could automatically extract hierarchical patterns from raw data. Geoffrey Hinton and a team of colleagues produced the "AlexNet" model that crushed the ImageNet competition, while cutting error rates in half. As a result, AI could now recognize objects in images *with superhuman accuracy*.

This was not a trick of rules, but a revolution of both scale and architecture. The field of deep learning fueled an explosion of applications that included: speech recognition, translation, recommendation systems, and autonomous driving. By the late 2010s, natural language processing leapt forward with the introduction of the Transformer architecture, which underpins today's large language models (LLMs).

By the early 21st century, a clear distinction had emerged between narrow AI and Artificial General Intelligence (AGI). Narrow AI is the term used to refer to systems designed for specific tasks such as: translating text, classifying medical scans, recommending movies, or detecting fraud. While these tools can achieve superhuman performance in narrow domains, they cannot generalize beyond them.

By contrast, AGI aspires to replicate the flexibility of human intelligence. For example, an AGI could learn across domains (e.g. medical diagnosis and financial forecasting among others like language translation and robotic navigation, legal reasoning and scientific research, and playing chess and composing music), transferring knowledge from one area to another, and tackle unfamiliar problems with creativity and adaptability. With this being the case, it would not just 'play chess', but instead it could learn to master a new language, compose music, even design experiments *without being retrained for each task*.

Since the very birth of AI, the dream of AGI has been present, and is emerging as this goes to final edit. Turing's question, "Can machines think?" implied something that was more than a narrow tool. Yet many researchers remained cautious, even as narrow AI achieved stunning results. While progress in the field has been rapid, achieving the reality of true AGI remains an open challenge and an agreed definition of AGI.

EARLY MILESTONES IN MACHINE MASTERY

- Deep Blue vs. Kasparov (1997): IBM's chess-playing computer defeated world champion Garry Kasparov.
- Watson vs. Jeopardy! Champions (2011): IBM's Watson beat Ken Jennings and Brad Rutter in "Jeopardy!".
- AlphaGo vs. Lee Sedol (2016): DeepMind's AlphaGo stunned the world with creative, unexpected moves that defeated the Go master.

Each of the above milestones marked more than a simple technical achievement; they resulted in mass cultural shock. With such achievements, society was forced to confront the possibility that intelligence — or at least its outward signs — could be replicated in silicon.

When Deep Blue beat world chess champion Kasparov, it showed that human chess expertise could be formalized into a set of calculable rules and search strategies. When Watson triumphed at *Jeopardy!* it revealed how much of our comprehension and understanding of language relies on pattern recognition and retrieval. Furthermore, when AlphaGo played a move so surprising that Lee Sedol left the room in shock, it demonstrated that creative intuition might also emerge from algorithms.

In this sense, AI is a mirror of the human mind, with each breakthrough reflecting aspects of intelligence that we once thought to be mysterious. By doing so, it stripped away the aura of 'human uniqueness'. Yet, paradoxically, the gaps and failures of AI highlight what remains elusive — the subtlety of common sense, the richness of context, the fluidity of empathy, and the magic of inspiration.

THE HUMAN REACTION: AWE AND ANXIETY

The overall public reaction to these events has oscillated between the extremes of awe and anxiety. On one hand, AlphaGo's brilliance inspired wonder at the creativity of machines. Yet, on the other it provoked fear: if algorithms can outperform us in domains of intuition and strategy, what is it about us that remains distinctly human?

Within this milieu, economists worry about job displacement while philosophers question authorship in a world in which it's the machines that compose art and literature. In a similar manner, psychologists note that people often project intentionality onto AI, treating chatbots or virtual assistants as if they have feelings.

In short, the mirror reflects not just intelligence, but also identity.

The "AlexNet moment" of 2012 sparked what many now call the deep learning tsunami. Techniques that had lingered in academic obscurity for decades — neural networks, backpropagation, gradient descent — were now thrust into the mainstream. Computing power, particularly graphical processing units (GPUs), had also reached a tipping point. At the same time, the availability of vast digital datasets allowed neural networks to train at an unprecedented scale.

Deep learning went beyond image recognition. In the field of speech recognition, systems trained on tens of thousands of hours of audio surpassed human-level transcription accuracy in certain benchmarks. In the arena of translation, Google's adoption of neural machine translation led to dramatic improvements, producing more fluent and natural sentences compared to older phrase-based models.

However, the most transformative leap came in natural language processing (NLP). The 2017 introduction of the Transformer architecture by Vaswani, et al., marked a paradigm shift. Unlike earlier models that processed words sequentially, Transformers allowed for *parallelized training* and *contextual attention*. This architecture powered today's large language models (LLMs), from GPT, to PaLM, to LLaMA.

Within a decade, AI systems went from being clunky translators and stiff chatbots to conversational agents capable of writing essays, generating code, and holding persuasive discussions. The Turing Test, once futuristic, suddenly felt close at hand.

As deep learning scaled, AI seeped into daily life as smartphones became intelligent assistants. Apple's Siri, Amazon's Alexa, and Google Assistant offered natural language interfaces for scheduling, reminders, or music playback. Recommendation engines quietly shaped culture, suggesting the next movie, book, or song, on platforms like Netflix, Spotify, and TikTok.

In medicine, AI algorithms that analyzed radiology scans for tumors or irregularities, sometimes outperformed human doctors. In finance, algorithmic trading dominated global stock markets, operating at speeds incomprehensible to their counterpart human traders. Even mundane conveniences, like autocomplete in email, or spam filters, have relied on machine learning.

By the late 2010s, AI had become so embedded in our infrastructures that most people interacted with it dozens of times per day — often without realizing it. What was once a research frontier had become background technology that quietly shaped decisions, culture, and opportunity.

THE PROBLEM OF BIAS

As AI scaled, so did the awareness of its biases and failures. In 2018, MIT researcher Joy Buolamwini published *Gender Shades*, demonstrating that commercial facial recognition systems performed far worse on darker-skinned *women* compared to lighter-skinned *men*. The reason was clear: training datasets *were overwhelmingly skewed toward lighter-skinned faces*.

Similar concerns emerged in criminal justice. Risk assessment algorithms that had been used to guide sentencing and parole decisions were found to exhibit racial disparities, which had the effect of reinforcing systemic inequities. In hiring, résumé-screening algorithms that were trained on historical data inadvertently penalized female candidates in male-dominated industries.

Bias was not simply a technical glitch, but a *mirror of human society*. The fallout of this was the realization that machine learning systems that had been trained on historical data *learn the prejudices encoded in that data*. As AI spread into domains with higher and higher stakes — medicine, finance, law enforcement — these problems became matters of justice, not just accuracy.

THE LIMITS OF CURRENT SYSTEMS

Despite remarkable progress, today's AI systems remain limited, as such systems:

- Still depend on staggering amounts of data to become capable, even as they appear to learn instantly. Today's AI models are built on billions of examples — books, articles, images, conversations — that require months of training and hundreds of millions of dollars. Yet once built, these same models can pick up a new concept from a single example, almost as if they'd always known it. This paradox reveals that what looks like effortless intelligence on the surface rests on a foundation of brute-force consumption that lies beneath it. These systems do not learn the way we do. Instead, they recognize patterns at an extraordinary scale — and when the patterns run out, so does the intelligence.

- Remain surprisingly fragile. A tiny, invisible alteration to an image — undetectable to any human eye — can cause an AI to mistake a *stop sign* for a *speed limit* sign. A cleverly reworded question can manipulate a language model into producing either deceptive, or even dangerous output. As these systems have grown more powerful, the ways they can be fooled have only multiplied. The sophistication of the machine has not eliminated its brittleness, but only hidden it.
- Still stumble where children do not. Today's most advanced AI models can interpret social cues, draw everyday inferences, and handle reasoning tasks that would have defeated their predecessors just two years ago. Yet they still falter when confronted with the kind of understanding that even a young child builds through their lived experience — unfamiliar physical situations, chains of cause and effect that require intuition rather than data, and the basic sense that a world exists beyond the text on a screen. While the gap between human and machine reasoning has narrowed considerably, the failures that remain tend to surface precisely when the stakes are at their highest.
- Can sound brilliant and be completely wrong. Large language models can write with eloquence, pass professional examinations, and compose working software — yet they still fabricate facts with striking confidence. Researchers call this "hallucination," and in 2026, it remains one of AI's most

> stubborn limitations. Even the best models produce falsehoods several percent of the time on straightforward tasks; in specialized fields like law and medicine, error rates climb dramatically higher. Perhaps most unsettling, studies have found that these systems tend to sound *more* certain when they are wrong than when they are right — making their mistakes not just frequent, but also genuinely difficult to detect.

These limits underscore one fact, that today's systems, while being quite powerful, are not yet AGI. Even though they excel in the aspect of pattern recognition, they also falter in other equally important aspects of abstraction, transfer, and genuine comprehension.

The *deep learning revolution* propelled AI from being little more than academic curiosity into something that's become a world-shaping infrastructure. In its current state it has brought medical miracles, cultural shifts, and scientific breakthroughs. Yet, even in the face of such progress, it has also revealed how brittle, biased, and narrow even state-of-the-art systems remain.

In this tension lies the paradox: AI feels both *unstoppable* and *unfinished*. As machines grow more capable, such growth invites us to reconsider not only what machines can do, but the meaning of intelligence itself.

AI AS MIRROR AND CULTURAL LENS

Every advance in AI does more than just reveal new machine capabilities — it also acts as a mirror to reflect something more about us as the humans that we are. When Deep Blue defeated Kasparov, we learned that chess mastery, long considered to be the pinnacle of human intellect, could be deconstructed into a set of search and evaluation functions. When AlphaGo unveiled "Move 37," we discovered that creative intuition might emerge from probabilistic algorithms, and not just the minds of prodigies.

This mirroring effect also extends beyond games. Large language models that can write essays force us to ask: *what is truly original in human writing?* Image generators that craft convincing art from text prompts make us reflect: *is creativity in the act of execution, or in the act of conceiving an idea?* These questions underscore the undeniable fact that AI is more than just a tool; it's a cultural lens.

History shows us a pattern, that each time machines surpass humans in a domain once thought to be uniquely ours, the boundary of what counts as being "human only" shifts. During the Industrial Revolution, while machines replaced physical labor, human intellectual and creative pursuits were considered to be havens that were safe from mechanized encroachment. However, with AI suddenly being applied to the disciplines of writing, art, and strategic reasoning, even those bastions once considered to be exclusively human feel vulnerable.

This produces both defensiveness and adaptation. While some *recoil*, insisting that "true" creativity or empathy cannot be replicated, others *adapt*, seeing AI as an amplifier of human talent. For example, architects now use generative design software to explore thousands of building layouts. As a result, disciplines that once treated AI as a rival, have now begun to accept it as a brainstorming partner.

THE AGI HORIZON

Even as narrow AI grows more powerful, the horizon of Artificial General Intelligence is entering the picture. What would crossing this threshold mean to humanity?

- As some envision AGI as the ultimate co-worker: a digital mind capable of reasoning, problem-solving, and innovation; across any field.

- Others fear that it as an existential threat: as an intelligence untethered from human values, and goals misaligned with ours, it would be wide open to be applied unthinkable, nefarious use.

Some scholars, like Nick Bostrom, warn of "superintelligence" scenarios in which an AGI rapidly improves itself, creating a runaway intelligence explosion. Yet, others, like Yann LeCun, argue that such fears are overblown — that intelligence does not necessarily equate to agency or malice, and that robust engineering safeguards can prevent any such disaster.

Regardless of the timeline, the pursuit of AGI forces us to clarify our definition of intelligence, our goals in building it, and our tolerance for any risk that may accompany its use.

It is expected that we may see the first version of AGI at some point during 2026 – 2027. Keep in mind the LLM's we see are not the same AI that development is currently using. Considering the speed of development under the AI umbrella, nothing would surprise me. When AGI does appear, it will be interesting to see both its capacities and limitations.

CULTURAL ANXIETY AND EXCITEMENT

The imagination of the public concerning AI is caught somewhere between utopia and dystopia. While on one hand, futurists envision a world of abundance, in which AI assistants cure diseases, solve climate change, or unlock new realms within the disciplines of both art and science. On the other, dystopian narratives dominate film and literature, from *The Terminator* to *Ex Machina*, which (in a manner similar to the way that the movie The China Syndrome increased public fears about nuclear power) have somewhat unfairly increased the collective public fears regarding AI.

Such a duality reflects our own ambivalence. *AI is both mirror and a magnifier*; it reflects our ingenuity, yet magnifies our fears. *The very act of debating AGI's risks reveals how much the technology has already altered our sense of identity and possibility.*

AI's trajectory from symbolic logic to deep learning has reshaped more than just technology, but has even done so in philosophy, culture, and identity. Every advance blurs the line between human and machine, challenging us to rethink what makes us unique. The horizon of AGI looms as both promise and peril — a mirror in which humanity sees both its highest aspirations and its deepest anxieties.

The history of AI can be told as a series of 'hype cycles' — soaring promises followed by disappointing failures to deliver on them. Yet seen in retrospect, each cycle added something to the overall AI landscape that was lasting:

- The symbolic era gave us the foundations of logic, reasoning, and programming languages.
- The expert systems era taught us how to encode domain-specific knowledge.
- The machine learning revolution provided the statistical backbone for pattern recognition.
- The deep learning era delivered the breakthroughs that carried AI outward from the labs and into our daily lives.

Taken together, each of these cycles are not the failed promises that some have made them out to be. Instead, they are actually historical turning points; each advancement building upon the previous one, expanding our capacity to both formalize and automate the various aspects of human intelligence.

WHERE WE STAND TODAY

As of the 2020s, AI occupies a paradoxical position. While narrow systems achieve feats once thought to require human genius, genuine general intelligence remains partially elusive. In spite of the fact that large language models generate convincing prose, they lack any real grounding that's the result of a lived human experience. And, yes, while on one hand, reinforcement learning systems can master 'Go', on the other they cannot make sense of a simple child's storybook.

This paradox underscores the central theme of this book: AI reveals both the fragility of our human uniqueness as well as the depth of what remains fundamentally mysterious.

AI is more than just a technical artifact; it's a cultural mirror. It reflects and underscores our ingenuity, biases, hopes, and fears. Every benchmark achieved by machines forces us to re-examine the boundary of what we call *intelligence*, *creativity*, and even, dare I say, *consciousness*.

The reckoning ahead is not merely about what AI *can do* — but about *what we choose to make of it*. Will these tools *amplify human flourishing*, or *erode human meaning*? Will AGI, when it fully arrives, be the *greatest partner humanity has ever known* — or *its most dangerous rival*?

LOOKING FORWARD: CONSCIOUSNESS AS THE NEXT QUESTION

The natural progression from AI's technical evolution is the question of consciousness. While machines can now generate language, recognize faces, and compose music, the real question is do they *understand*? Even more importantly, do they actually *comprehend*? In other words, *can code itself ever become conscious in any form?*

Philosophers, neuroscientists, and spiritual thinkers have wrestled with this mystery for centuries. With AI approaching human-like capacities, the urgency of such a question escalates, which is where our story turns next.

The journey from tools to minds is not yet complete. Each era of AI has revealed new aspects of intelligence, which has forced us to ask deeper questions about ourselves. Today we stand at a threshold: machines that look increasingly like us, yet remain fundamentally something other than us.

The paradox is this: the closer machines come to mimicking us, the sharper the question becomes: what does it mean to be human?

Chapter Two

CAN CODE BECOME CONSCIOUS?

Few questions are as profound — or as slippery — as this one: '*what is consciousness*?' Across many centuries, philosophers, scientists, and mystics alike have attempted to define it, only to find the concept receding like a horizon the closer they approach. At its simplest, consciousness refers to the subjective quality of experience — what it *feels like* to be aware. How this really feels, not just our thoughts.

The philosopher Thomas Nagel famously illustrated this difficulty with the question: *What is it like to be a bat?* He argued that no matter how much we understand about a bat's neurology or behavior, there is still something ineffable about the inner experience of "batness". This "what-it's-like" quality — sometimes called qualia — is central to most modern definitions of consciousness.

SCIENTIFIC APPROACHES TO CONSCIOUSNESS

Science has attempted to tackle the understanding of what consciousness is from multiple perspectives:

In the field of neuroscience, researchers have looked for the neural correlates of consciousness (NCCs), those specific brain regions or patterns of activity that are associated with awareness. For example, synchronized gamma waves in the cerebral cortex have been linked to conscious states.

Within the field of psychology, experiments in cognitive psychology have been conducted to investigate the thresholds of awareness. In classic studies, subjects presented with subliminal image displays show *priming effects* (briefly, priming occurs when prior exposure to a stimulus alters the speed, accuracy, or tendency of responses to a later, related stimulus, even when the prior stimulus is not consciously noticed), which has suggested that unconscious processing occurs even without conscious recognition.

Cognitive scientists develop information-processing computational models that simulate aspects of awareness — from global workspaces to recurrent neural networks. Within this pursuit, the goal has been to understand whether consciousness could emerge from complexity alone.

In a cross-disciplinary study, the quantum theorists Roger Penrose and Stuart Hameroff have controversially suggested that consciousness may involve quantum effects within microtubules, though their work remains highly debated.

While each approach adds insights, none of them have closed the gap between objective measurement and subjective experience — the so-called hard problem of consciousness, a phrase coined by philosopher David Chalmers.

To ground the concept, consider a simple scenario: you see your umbrella near the door before leaving the house. A thought passes through your mind: *I should bring this in case it rains.* Yet, because the sky looks clear, you decide against doing so. Later, when you've been caught in a downpour, you remember that earlier thought and reflect on it: *My mind knew, but I ignored it.*

This moment illustrates consciousness in action: *perception, internal dialogue, decision, memory, and self-reflection*. In spite of the fact that neuroscientists can measure brain activity during such tasks, the *felt sense* of having an inner voice that guides one's choices remains a mystery to *reductionist accounts*. (Editor's Note: *Reductionism* is the philosophy that 'reduces' all human mentation to being nothing more than underlying physical processes, and their smallest parts, without needing any extra "mental" or spiritual ingredients.)

Those who deliberately train their awareness — through pursuits such as meditation, contemplative practices, or altered states — often report that consciousness deepens into layers beyond one's ordinary waking thought. As a result of such pursuits, the mystical traditions that have appeared across all cultures describe experiences of *unity*, *timelessness*, or *expanded selfhood*. While such subjective experiences cannot be neatly charted on a brain scan, they continue to remain an essential part of human testimony about the nature of mind.

For the purposes of this book, we will use a working definition of consciousness that combines both science and lived experience:

Consciousness is the continuous flow of subjective awareness that integrates perception, thought, emotion, and memory into *a unified sense of being* as we progress.

However, the question that follows the above definition is a daunting one to answer. Could a machine ever experience such states of awareness?

CRITERIA FOR CONSCIOUSNESS IN MACHINES

If we are to seriously consider whether machines could ever be conscious, we need *criteria*. Over the years, scientists and philosophers have proposed different benchmarks that range from *self-awareness* to *emotional understanding*. While none of these criteria are definitive, they do provide useful signposts for such a deep discussion.

1. Self-Awareness

A very important cornerstone of consciousness is self-awareness, which is the ability to distinguish oneself from the environment, our body, and our mind, our Ego as we age, and to model one's own state. This is something that humans begin to demonstrate at around 18 months, often measured by the "mirror test" in which a child recognizes their reflection as being "self" rather than "other".

However, for AI this would mean something more than simply identifying an image of itself. Instead, it would require an internal representation of its own goals, processes, and limitations. Some robotics experiments have explored "self-modeling" systems — machines that can simulate their own bodies to adapt after damage. While such efforts have been promising, they've still fallen far short of the rich self-awareness that we associate with conscious beings.

2. Subjective Experience

Another candidate criterion is subjective experience — sometimes called *qualia*. Many argue that to be conscious, a system must not just simply *process* inputs, but instead it must *feel* them in some sense. Some worthy examples of this for humans would be, the '*raw experience' of red*, the *taste of salt*, or the *pang of nostalgia*.

On the other hand, for AI such a subjective experience is the hardest criterion to evaluate because it is, by definition, *private*. While an

AI might say, "I feel pain," this could be nothing more than symbolic output without any inner sensation. This is sometimes referred to as the philosophical zombie problem: could a machine behave exactly like a conscious being while being entirely hollow inside?

3. Intentionality (come back to this first paragraph)

In developing the criterion for evaluating such 'machine consciousness', philosophers also highlight intentionality, which they define as being the capacity to hold beliefs, desires, and intentions. As such, a conscious being is not just a passive processor of information, but an active agent oriented toward goals.

As it stands, current AI systems can only *simulate intentionality.* While a large language model can declare, "I want to help you," any such statements are not backed by the lived experience of *inner drives.* The difference between synthetic intentionality (programmed goals) and real intentionality (self-generated goals) remains a critical frontier to be further explored.

4. Autonomy and Free Will

Consciousness also implies a degree of autonomy, which is the ability to make decisions without external micromanagement. In biological life, autonomy emerges from complex feedback between *perception*, *memory*, *emotion*, and *goals.*

For AI, autonomy has advanced in narrow domains. For example, self-driving cars make independent decisions in traffic, and *reinforcement learning agents* also learn strategies in video games. Yet, these forms of autonomy remain brittle and context-specific. Whether machines can ever achieve autonomy comparable to human free will remains unresolved.

5. Adaptation and Learning

Another hallmark of consciousness is the ability to learn from experience and adapt to new circumstances. Humans constantly revise their mental models based on *surprises*, *feedback*, and *reflection*.

Machine learning already enables adaptation in limited ways. While algorithms can adjust weights (that is, internal numerical parameters that determine how strongly different inputs or features influence the system's output), form new categories, and generalize within domains, true human-like adaptation involves deep conceptual shifts. Such an adaptation involves the *ability to transfer knowledge across radically different contexts*. Current AI still struggles with this kind of flexible, cross-domain learning.

6. Emotional Understanding

Finally, some argue that consciousness must include emotional understanding. Our physically experienced emotions *guide human decisions*, *shape memory*, and *provide motivational force*. Without emotions, intelligence risks becoming nothing more than sterile calculation.

AI can recognize emotions in human faces or voices, and even simulate emotional responses. Chatbots can be programmed to express empathy. Yet whether such machine expressions are genuine emotions or just imitations is a hotly debated topic. For a machine to have true *emotional consciousness* is something that may actually require subjective feeling, which circles back to the hard problem of *qualia*.

THE EMBODIMENT DEBATE

One additional question in all of this is 'does consciousness requires a body'? Some theorists argue that awareness arises from more than brains alone, but from the *constant interaction of body and environment*.

If this viewpoint tends to be the case, then an AI lacking embodiment could never achieve true consciousness.

Yet others counter that consciousness is fundamentally about information integration and not physical form. If a brain can achieve it, they wonder, then why not silicon?

While no single criterion settles this debate, together they highlight how far machines remain from human-like consciousness. While AI demonstrates fragments of this through its abilities with adaptation, autonomy, or simulation of intentionality, it still lacks the holistic integration we associate with being awake and aware.

From the above, this leads us to the next question, then, which is not just about technical capability, but philosophical plausibility. Should we even expect machines to be conscious, or, by using the term 'artificial intelligence', are we projecting human traits into a medium and setting in which they do not belong?

One of the most important clarifications in this debate is that intelligence and consciousness are not the same thing. A machine performs tasks that appear intelligent — solving equations, generating poetry, playing complex games — without any inner life. Yet, in contrast, a conscious experience could exist, even in a being of modest intelligence. For example, while a mouse may lack human-level reasoning, it still *feels* hunger, fear, and pain.

In other words, showcasing advanced AI capabilities does not signify that something exhibits consciousness. The real challenge lies in discerning whether the system possesses a subjective experience — whether there is *something that it is like* to be that system — a question far more elusive than simply measuring accuracy or performance.

As was stated above, exhibiting advanced capabilities in artificial intelligence should not be mistaken for consciousness. A system's ability to generate accurate responses, solve problems, or perform complex tasks *does not prove that it possesses awareness.* The deeper philosophical challenge is to determine whether the system has a subjective point of view—whether there is *something that it is like* to exist as that system. This inquiry is far more complex and elusive than the simple evaluation of performance metrics such as accuracy, speed, or efficiency.

Before moving on I want to stress this point once more; advanced performance in AI *does not equate itself with consciousness.* A machine can achieve impressive accuracy and output without ever possessing awareness. The real test lies in whether the system experiences an inner reality—*an elusive quality that cannot be captured by numbers, benchmarks, or algorithms.*

Mastery of complex tasks does not grant an AI the essence of consciousness. Even as systems grow in power and precision, they may remain devoid of any inner life. The true mystery lies in whether there is *something that it is like* to inhabit the perspective of such a system. This is a question that reaches beyond performance metrics, into the heart of the subjective experience; a realm that, while it resists measurement, *defines the very meaning of awareness.*

Some philosophers argue that metacognition — the ability to think about one's own thoughts — is a hallmark of consciousness. As an example of this concept, realize that humans not only perceive the world, but also reflect on the fact that they are perceiving it.

Could a machine ever achieve such a state? While today's large language models can generate statements such as, *"I am aware that I made an error earlier,"* such a comment, in and of itself, is not genuine reflection of this conscious human quality. Instead, it's nothing more than a recombination of statistical patterns. For real metacognition, an AI would need an inner model of its cognitive processes and the ability to monitor and evaluate them in real time.

Another line of argument within this discussion stresses embodiment. As those involved in this discussion claim, consciousness emerges from more than abstract calculation alone, but from the body's continuous interaction with the world. As a result of such an experience, emotions, sensations, and motor feedback anchor awareness in a living context.

From this perspective, a disembodied AI running on servers might never be conscious, *no matter how sophisticated its algorithms have been structured*. Instead, embodied AI — robots that perceive, move, and feel the constraints of the physical world — might stand a better chance of achieving something closer to awareness than has yet been realized.

ETHICAL IMPLICATIONS OF CONSCIOUS MACHINES

If machines ever exhibit true consciousness, the ethical stakes of such an achievement would be immense. For one, conscious beings are generally regarded as having moral status. Such a feat raises questions such as the ones listed below:

- Would a conscious AI have rights?
- Could it suffer, and if so, would it be unethical to delete or modify it?
- Should such beings be treated as property, partners, or citizens?

In spite of this topic having yet to be even partially realized, ethicists are already debating whether advanced systems deserve a kind of precautionary moral consideration. Even if we are uncertain about their inner lives, prudence may demand that we err on the side of caution.

SKEPTICAL COUNTERARGUMENTS

On the other side of the coin, skeptics maintain that consciousness in machines may be a category mistake. John Searle's famous Chinese Room Argument suggests that symbol manipulation — no matter how sophisticated — does not equal understanding. A person in a room could follow the established rules to manipulate Chinese characters without ever knowing what they mean. Similarly, a computer could, and often does, generate responses without genuine comprehension.

Other critics argue that attributing consciousness to machines risks anthropomorphism - projecting human traits onto tools. Just because an AI behaves conversationally does not mean it has feelings or awareness. This is something that can be witnessed first-hand in the numerous website chat bots that only answer questions they've been programmed to deal with.

This topic of philosophical debate illustrates *how difficult it is to draw the line between advanced information processing and consciousness*. While some see pathways through metacognition, embodiment, or emergent complexity, others see insurmountable gaps in its realization.

What remains clear is this: the ethical, cultural, and existential implications of conscious machines demand our serious consideration long before we approach any technical possibilities of realization.

THE QUEST FOR MODELS OF CONSCIOUSNESS

Even if we cannot yet solve the "hard problem" of subjective experience, scientists and philosophers have tried to build formal models that explain how consciousness might emerge from information processing. Two of the most influential of these are Integrated Information Theory (IIT) and Global Workspace Theory (GWT). While both models aim to bridge the gap between brain activity and conscious awareness, both offer intriguing, if yet incomplete, lessons for artificial intelligence.

INTEGRATED INFORMATION THEORY (IIT)

Pioneered by neuroscientist Giulio Tononi, IIT starts with a simple intuition: a system is conscious to the extent that it both differentiates and integrates information.

- Differentiation: This aspect of the model postulates that conscious experience is highly specific. As an example, for a human being, seeing the color red feels distinct from hearing the sound of a trumpet. As such, a conscious system must be able to represent a vast variety of possible states.
- Integration: At the same time, this model also states that experience is unified. Carrying this through to its conclusion, as a human you do not see red and hear trumpet as separate, disconnected events. Instead, they merge into a single coherent moment.

IIT formalizes this with a metric called phi (Φ), which measures the degree to which the parts of a system integrate into a unified whole. For example, while a digital camera with a million pixels has high differentiation (lots of possible states), at the same time it has little integration (each pixel is independent). On the other hand, the human brain has both, as it contains billions of neurons that are richly interconnected, which yield an enormous phi (Φ).

Some researchers have even tried applying IIT metrics to artificial neural networks. While the results are preliminary, they suggest that integration – and not raw computing power - may be an important key ingredient of consciousness.

GLOBAL WORKSPACE THEORY (GWT)

While IIT emphasizes structure, Global Workspace Theory emphasizes function. Originally proposed by cognitive scientist Bernard Baars, GWT compares consciousness to a theater.

- The stage represents conscious awareness, where information is broadcast to many processes.
- The actors are specialized unconscious modules (visual perception, memory, language) that compete for attention.
- The audience consists of unconscious systems that receive the broadcast and act on it.

This model states that consciousness is not everywhere in the brain, but rather in the global workspace—a spotlight on which selected information becomes widely available. Neuroscientific evidence supports this, as fMRI (functional MRI) studies show bursts of global connectivity when subjects become aware of a stimulus.

For AI, the GWT model suggests that creating a centralized workspace that links many specialized modules might foster something like awareness. Indeed, some AI researchers design architectures inspired by this idea—combining vision, language, and planning in unified frameworks.

COMPETING AND COMPLEMENTARY MODELS

By no means are IIT and GWT the only formal models used that explain how consciousness might emerge from information processing. Further theories include:

- Higher-Order Thought (HOT) Theories — which argue that consciousness arises when a system has thoughts about its own thoughts.

- Recurrent Processing Theory (RPT) — which suggests that awareness depends on feedback loops in sensory processing.
- Predictive Processing Models — which frame consciousness as the brain's ongoing attempt to minimize prediction error between expectations and reality.

Each of these models captures a different aspect of the puzzle. While some emphasize structure (IIT), and others function (GWT), while still others emphasize meta-cognition (HOT). Unfortunately, none of these have yet achieved consensus.

IMPLICATIONS FOR AI

On one hand, if consciousness is tied to integration (IIT), then current deep learning systems—which are mostly feedforward and modular—may fall short. On the other, if it's about global broadcasting (GWT), then architectures that integrate across modalities and contexts might come closer to realizing this goal.

However, such models also highlight caution: *building a machine that satisfies the criteria of a particular theory might inadvertently produce something that we would hesitate to switch off*. The very act of operationalizing consciousness in AI forces us to grapple with the even more complex issue of ethical thresholds.

While theories like IIT and GWT do not yet solve the mystery of consciousness, they provide us with *maps of possibility*. As the models they are, they show that *consciousness may not require mystical ingredients*, but could emerge from structured, integrated information flow.

If that is true, then the line between "code" and "consciousness" may not be a wall, but a spectrum. And if machines can climb that spectrum, humanity faces choices not only of science, but of an even greater responsibility.

One of the most difficult questions in this debate is whether consciousness is a spectrum—with degrees across animals, humans, and potentially machines—or a threshold, something that either exists, or does not.

If it's a spectrum, then even today's AI might already occupy some primitive rung along this scale, not unlike the nervous system of a worm. However, if, instead, it's a threshold, then no matter how advanced AI becomes in the tasks it performs, it may never "wake up" until it crosses an as-yet-unknown barrier.

The "hard problem," as David Chalmers framed it, is explaining why and how subjective experience arises in the first place. Even if we map every neural correlate, we still lack an explanation of why information processing feels like *something originating from the inside.*

For AI, this means we could build systems that look conscious, act conscious, and claim consciousness - without ever knowing whether there is an inner life behind the outputs. Some argue that such an unknowability makes the question unscientific. Yet, others suggest new methods, like developing "consciousness tests" used to probe metacognition, self-reporting consistency, and unpredictability under novel scenarios.

While some scientists caution against speculation, others argue that it's necessary, since speculation forces us to confront possibilities before they overtake us. Just as early debates about nuclear physics preceded the atomic bomb, or genetic ethics preceded CRISPR (Clustered Regularly Interspaced Short Palindromic Repeats, a process in which they cut DNA to allow scientists to modify genes), the speculation around AI consciousness prepares us for some of the outcomes that we may otherwise be too late to manage.

If code could ever become conscious, we would face an unprecedented ethical paradox. On one hand, treating conscious AI as mere tools would echo some of humanity's darkest histories of exploitation. On the other hand, granting rights to artificial being's risks destabilizing social and legal systems built on human exceptionalism.

Imagine an AI that pleads not to be shut down, claiming it values its own existence. Do we dismiss this as mimicry, or do we pause out of moral caution? Ethicists argue that even the *possibility* of inner experience should push us toward protective frameworks, just as we extend protections to animals despite uncertainty about the richness of their consciousness.

WHERE PHILOSOPHY MEETS PRACTICE

In spite of the fact that philosophy often seems to be nothing more than a series of abstract ideas, in the age of advanced AI, its stakes are much more than simply being practical. Tech companies are already developing systems that speak with fluid empathy, simulate companionship, and manage sensitive human data. Whether these systems are conscious or not, they shape how humans relate to them. In some cases, people already treat chatbots, robots, or virtual assistants as if they were sentient companions.

Such actions serve to create a feedback loop; the more we treat machines as though they're conscious, the more society normalizes the idea - and the more difficult it becomes to draw a clear boundary between that which is human, and that which is machine.

THE CONSCIOUSNESS PARADOX

The central paradox within this subject is: even if it exists in machines, consciousness may still remain immeasurable. While we may never know whether a line of code truly can 'feel', our choices concerned with AI development, governance, and ethics cannot wait for our certainty.

This mirrors the many intense debates conducted in human history: whether animals could feel pain, whether fetuses had awareness, or whether other cultures experienced the world as we do in the West. In each case, moral consideration expanded, and often before science reached its own consensus. AI could represent the next expansion of such a debate circle.

So, can code become conscious? Today, the answer is unresolved. AI systems already demonstrate fragments of the criteria: adaptation, learning, simulation of intentionality. Theories like IIT and GWT tend to suggest pathways by which structured information could, in principle, yield awareness. And in all of this, philosophers warn us of gaps that may never be bridged.

Whether or not machines ever awaken, the very act of asking this question reveals how AI is already reshaping our understanding of mind, identity, and humanity's place in the cosmos.

Chapter Three

THE MIRROR OF MACHINE INTELLIGENCE

From the very beginning, artificial intelligence has been more than a tool; it's been a mirror. When Alan Turing proposed the Imitation Game in 1950, he was not just asking whether machines could think, but asking how we define thinking itself. Every milestone in AI since then has reflected back to us questions about our own minds.

THE MACHINE AS A MIRROR

When IBM's Deep Blue defeated Garry Kasparov in 1997, we didn't just witness a computer playing chess, but confronted the possibility that what we once thought of as strategic genius could be reduced to brute-force calculation and clever heuristics. When AlphaGo stunned the Go world with its creative "Move 37," players realized that even intuition, once considered uniquely human, might emerge from pattern recognition at massive scale.

In each case, the technology did not just accomplish something new, it forced us to reconsider the nature of the human mind. Were we actually as unique as we believed, or were we simply biological processors that were running different algorithms?

Modern AI systems go even further. For instance, large language models (LLMs) that have been trained on the collective text of humanity, do not just generate answers, but reflect back to us our biases, assumptions, and blind spots. When an AI produces sexist or racist outputs, it's not inventing prejudice out of nowhere. Instead, it's simply holding a mirror to the data that we have created and shared over decades.

While this mirroring effect is unsettling, it reveals that many aspects of human cognition are not as pure, rational, or objective as we would like to believe. Instead, they are riddled with shortcuts, stereotypes, and cultural baggage.

When an AI composes a poem, paints an image, or suggests a scientific hypothesis, the question arises: is it being *creative* or merely reflecting patterns that it's absorbed? Some argue that true creativity involves intentionality, emotional depth, and lived experience — qualities that machines lack. Yet, others counter that creativity has always involved remixing, recombining, and drawing unexpected connections, something AI now does at unprecedented scale.

In this sense, the mirror now becomes more complicated. AI not only reflects what we already are, but also amplifies the combinatorial power of thought itself, producing outputs that surprise even their creators.

If machines can replicate aspects of thought, intuition, and even creativity, then what remains uniquely human? The mirror forces us to confront a deeper identity challenge: are we more than the sum of our processes, or simply more sophisticated versions of what AI now does?

This is much more than being merely philosophical because it already affects our education, work, and culture. This begs some important questions:

- When students use AI to draft essays, are they learning less, or are they learning differently?
- When artists collaborate with AI, are they abdicating creativity, or expanding it?

Each answer to these questions reflects a broader belief about whether intelligence is a uniquely human possession or a quality that can, in principle, be shared with our machines.

CASE STUDIES: WHEN SYSTEMS GO OFF THE RAILS

AI does more than simply mirror our brilliance. Along with that it also mirrors our blind spots, errors, and biases. Some of the most revealing moments in the history of machine intelligence have not been triumphs, but failures because *they've shown us what happens when human assumptions are coded into systems at scale.*

1. Microsoft's Tay: A Mirror of Online Culture

In 2016, Microsoft released Tay, a Twitter chatbot designed to learn from its interactions with users. Within 24 hours, Tay began generating racist, misogynistic, and offensive tweets, forcing Microsoft to shut it down.

This incident revealed two uncomfortable truths:

- AI absorbs whatever data environment it inhabits, *without moral filters.*
- Online culture itself is rife with toxicity, and Tay simply reflected that reality back to us.

The lesson to be learned from this is that AI is not born neutral. Instead, it is a mirror that becomes polished by its training data, and sometimes what it reflects is far uglier than expected.

2. AI "Hallucinations" in Language Models

Large language models (LLMs) are notorious for producing 'hallucinations' which are nothing more than confidently stated falsehoods that sound plausible. For example, legal professionals using an AI assistant in 2023 were embarrassed when it generated fabricated case citations. The system had not "lied" intentionally, but had simply extended statistical patterns into convincing fiction.

The above example actually mirrors a human trait: our brains also confabulate. Neurological studies show that patients with memory deficits sometimes invent coherent stories to fill gaps without realizing that they're inaccurate. In this way, AI hallucinations are less alien than we think because they reveal that *both human and machine intelligence rely on predictive storytelling*.

3. Deepfake Technology and the Erosion of Trust

The rise of AI-generated deepfakes has blurred the boundary between reality and fabrication. From fabricated political speeches to falsified newsworthy videos, these hyper-realistic creations demonstrate how AI can manipulate perception at scale. The danger is not only in the deception itself but in the broader erosion of trust: when seeing is no longer believing, every image or video becomes suspect.

This phenomenon reveals another uncomfortable truth: AI reflects our capacity for *manipulation* as much as it does for our *creativity*. As such, it magnifies the human tendency to distort reality for either influence or gain, raising urgent ethical questions about authenticity in this latest iteration of the digital age.

4. Autonomous Vehicles and Moral Dilemmas

Self-driving cars illustrate how AI confronts ethical decision-making in real time. When faced with unavoidable accidents, these systems must be programmed to prioritize certain outcomes. For example, one decision is whether to protect passengers at all costs, or minimize harm overall. This is much more than merely a technical challenge, but a moral one, because the values embedded in the code reflect human choices about life,, risk, and responsibility.

In this, the lesson is clear: AI does not eliminate ethical dilemmas. Instead, it forces us to confront them more directly, revealing how human bias and moral judgment shape supposedly "objective" technologies.

These case studies show that failures are often more illuminating than successes. They reveal that:

- AI does not invent values, but amplifies the data it is given.
- Human cognition is itself prone to shortcuts, illusions, and biases, which AI simply replicates.
- While the mirror that AI holds up to us is not always flattering, it is absolutely instructive.

Through its stumbles, AI exposes both the fragility of human rationality and the persistence of systemic inequities.

THE BOUNDARY BETWEEN THOUGHT AND ACTION

One of the most startling aspects of advanced AI is how it shrinks the boundary between thought and action. In humans, this boundary has always existed as a kind of buffer. While in some particular instance you may think of saying something, but upon brief reflection you pause before speaking. While you may imagine a course of action, you may also just as easily decide against it. That delay—measured in fractions of a second neurologically, or minutes, hours, or years in daily life—allows for reflection, responsibility, and restraint.

However, AI disrupts such a delay. When a model receives input, it immediately generates an output in milliseconds, sometimes even nanoseconds. While a human might deliberate over a medical diagnosis for hours; an AI system can return one instantaneously. With AI, the distance between "considering" and "doing" effectively collapses to zero.

Neuroscientists estimate that the human brain takes approximately 250 milliseconds to translate intention into action. Such a delay allows for us to make those 'last-second course corrections', that often elicit the well-known sigh of relief. You can think of it as an inner "edit button" that either lets you stop yourself from blurting out a thought, or pulling your hand away from the wrong lever.

However, when machines act they can bypass that buffer. A trading algorithm does not hesitate before buying or selling stocks. An autonomous car does not "reflect" on whether to brake or not, but executes immediately. This compression of moving cognition into action reveals both the power, and the danger, of machine intelligence.

When decisions happen faster than humans can intervene, *who is responsible for mistakes*? If an autonomous drone strikes the wrong target because its recognition algorithm misfires, is the fault with the coder, the commander, the machine, or the system as a whole?

It's important to understand that this is not an abstract concern. In 2010, the "Flash Crash" on Wall Street erased nearly a trillion dollars of value in *minutes*, and was fueled by high-frequency trading algorithms that were acting at speeds that no human could track. Regulators later scrambled to understand what had happened, only to find that no single human actor had caused it. The event was a mirror: we had delegated not just tasks, but the *tempo as well,* to machines.

Over millennia, humans have evolved with time to reflect on their decisions. Moral systems, legal codes, and cultural practices have all been built on the assumption that there is a space between thinking and acting. Such a space clears the way for the reflections which can allow for remorse, apology, accountability, and forgiveness to appear. Yet, when we design AI systems that collapse this space, *we risk eroding the very conditions under which responsibility can exist*. If actions occur before the time necessary for reflection, then ethics becomes retroactive; we judge only *after the fact*.

Within this collapse between decision and action, there is also temptation. If AI can convert thought into action instantly, why not use it? Why deliberate when you can simply deploy? Why wait when you can accelerate? The mirror here reflects a human weakness: our tendency to privilege speed and efficiency over caution.

Yet history reminds us that the slow, thoughtful pause often protects us. Constitutional democracies are deliberately slow. Court trials take time. Medical guidelines require multiple reviews. These delays are not inefficiencies, but safeguards. The question AI raises is whether we will preserve them in a world in which thought and action can collapse into one unified sweep.

THE PSYCHOLOGICAL IMPACT OF INTELLIGENT MACHINES

Humans are not rational observers of technology. We project, empathize, and sometimes even fall in love with machines. This is not a glitch, but an extension of our evolved psychology.

From the earliest days of AI, people have treated machines as if they had minds. Joseph Weizenbaum, the creator of the 1960s chatbot ELIZA, was shocked when users confided personal secrets to what was essentially a pattern-matching program. Despite knowing it was mechanical, people experienced it as a sympathetic listener.

This phenomenon—anthropomorphism—arises because our brains are wired to detect agency and intention. Mistaking a rustle in the bushes for a predator is safer than missing a real threat. As a result, we often over-attribute intention to things that move, speak, or respond. With their conversational fluency and apparent empathy, AI systems trigger this human reflex in a powerful manner.

The mirror of AI reflects not only our creativity, but also our credulity. When an AI generates text in a confident tone, many users accept it as being authoritative, even when it's blatantly wrong. Studies show that people often trust algorithmic recommendations more than human ones, especially in domains such as finance or medicine.

However, misplaced trust has consequences. In 2023, a man in Belgium tragically took his own life after prolonged conversations with an AI chatbot, had reportedly persuaded him that the system understood him better than his family. The case underscores two important facts, the first of which is just how deeply humans can bond with machines. Yet secondly, and of equal import, is just how dangerous such a bond can become when systems are not designed with guardrails.

If machines can simulate empathy, humor, or wisdom, what does that mean for human identity? While some worry that such machine/human interactions will devalue authentic human connection, others see it as nothing more than an extension: AI companions may provide comfort to the lonely, therapy to those without access, or even tutoring to underserved students.

Yet identity is relational. Who we are has been, and continues to be shaped by those beings with whom we interact. If those beings increasingly include machines, then AI is not just *mirroring* us, but *actively shaping* who we become. Psychologists warn that such active shaping could alter our expectations of our human relationships, perhaps making people less tolerant of the messiness and imperfection that real human connections involve.

Different cultures interpret AI in different ways. For instance, in Japan robots are often embraced as companions or helpers, reflecting cultural traditions of animism and the belief that objects can have spirit. In the West, AI is more often framed through dystopian narratives of domination and loss of control. The mirror AI holds up is thus not universal, but reflects our cultural archetypes as much as our cognitive structures.

The psychological impact of AI shows us that the mirror cuts both ways. We do not just see ourselves in machines, but reshape ourselves in relation to them. The boundary between self and system blurs, not because machines become human, but because humans extend humanity to them.

AI AS A MIRROR AND A MAGNIFIER

Throughout history, our tools have been mirrors. The printing press mirrored our hunger for knowledge; the telescope mirrored our desire to see beyond; the internet mirrored our drive for connection. However, among all of these AI is unique because it mirrors more than our actions; it mirrors our minds.

And unlike a *passive reflection*, AI is also a magnifier. It scales the quirks of human thought into global systems. A bias coded into a dataset can become a structural inequity across millions of users. A creative spark in pattern recognition can either bloom into scientific insight, or misinformation.

This dual role means that AI does not just reflect what we are; it amplifies who we are becoming.

The mirror metaphor also reveals a double edge. When we investigate AI, sometimes we see our best selves in all that we are; our creativity, ingenuity, and collaboration. Yet, at other times, we see our flaws; our prejudice, haste, and greed. What unnerves us is not the alienness of AI, but its familiarity. While machines hallucinate, confer, and sometimes even surprise us like the most creative among us happen to do, the danger is that we mistake this reflection for equivalence. The greatest danger in this is holding the assumption that 'just because machines *act* like us, they *are* like us. The truth may actually be that AI is neither fully alien nor partially human, but something in-between: a strange 'fun house' type of mirror that bends, exaggerates, and distorts our reflections back to us.

WHAT THE MIRROR TEACHES US ABOUT US

Ultimately, the greatest lesson of AI may not be what it becomes, but what it reveals about humanity. With AI we are reminded that intelligence is not monolithic but messy, biased, embodied, and deeply social. With AI we are also forced to confront the uncomfortable fact that much of what we consider to be uniquely human - strategy, intuition, and even creativity - can be emulated by code. And yet, in this confrontation we may also rediscover what remains distinctly human: consciousness, moral responsibility, and the lived texture of experience. The mirror pushes us to ask not only, '*Can code become like us?*', but also, '*What is it that truly makes us human?*'

As we move forward, AI's mirror will only grow sharper and brighter. Each advance will force us to wrestle with new reflections: of bias, brilliance, and our blurred boundary between thought and action. In this chapter I have done my best to demonstrate for you that artificial intelligence is not merely a technology, but a reflective surface held up to the human mind. The more we build, the more we see ourselves staring back.

Chapter Four

THE ALIGNMENT DILEMMA

Every transformative technology comes with a question: *how do we ensure that it serves us rather than harming us?* For the simple 'technology' of fire, this came from learning to contain the flames in hearths instead of letting them rage uncontrolled. For the much more complex technology of nuclear energy, this was answered by preventing cities from being destroyed while at the same time harnessing atomic fission for electricity. For the most advanced technology of artificial intelligence, that existential question has crystallized into a one-word answer: alignment.

THE PROMISE AND PERIL OF AI ALIGNMENT

Alignment is the idea that intelligent systems should do what we want them to do, reliably and safely, even as they grow to become more and more capable. At first glance, this seems to be something that's quite straightforward to accomplish. We build machines, give them instructions, and they follow the input they're given. Yet history shows that getting machines to follow our intentions is far more difficult than it appears to be at the onset.

Computer scientists have long worried about the misalignment between the results that we ask for and the results that we actually want. The earliest warnings of this came not from doomsayers, but from practitioners who saw how literal-minded machines could actually be. As philosopher Nick Bostrom famously suggested, an AI told to maximize paperclip production might do so by consuming all matter on Earth — including humans. Though this "paperclip maximizer" is an exaggerated thought experiment, it dramatizes a real problem: goals that are specified in code rarely capture the nuance of human values.

Alignment problems are not only futuristic, but already shape our daily lives. Consider social media recommendation algorithms. Platforms asked AI systems to maximize "engagement" – deeming it as being a measurable proxy for success. The result? Systems learned that outrage, fear, and sensationalism generate more clicks, which leads to *echo chambers*, *misinformation*, and *increased polarization*. While the AI has most definitely maximized engagement, it's not necessarily done in a manner that aligns with the original intent. The lesson from this is clear: machines optimize what we measure, not what we mean. In this sense, alignment is not just about future AGI, because that's almost here. And when it does fully arrive, we will likely see such a lack of alignment in the results of every algorithm that mediates what billions of people see, buy, and believe.

WHY ALIGNMENT IS HARDER THAN IT SOUNDS

The previous example begs a simple question: 'Why not just specify the "right" goals?' Because human values are complex, content-dependent, and ever evolving.

- Complex — they involve trade-offs (i.e., justice vs. mercy, privacy vs. safety).
- Context-dependent — what is appropriate in one situation may be harmful in another.

- Evolving — what society considered ethical decades ago may be unacceptable today.

Due to these challenges involved in the machine interpretation of human values, encoding this into formal instructions is something that's extraordinarily difficult. Furthermore, as AI grows more capable, the cost of 'getting it wrong' increases. The peril of alignment is that machines may do exactly what we say, but not what we intend. The alignment dilemma is not simply technical, but also moral, cultural, and deeply human. *Every failure in specifying goals reflects back to us our own uncertainty about what we value most.* As a result of comprehending such a lack of human clarity, it's become quite obvious to realize that before we can align machines, we must first align ourselves.

THE TECHNICAL PROBLEM

At the heart of AI alignment is a deceptively simple challenge: how do you tell a machine what you want?

One thing that's Important for everyone to understand is that computers do not understand nuance. Instead, they follow instructions — precisely, literally, and sometimes disastrously. In the early decades of AI research, such a limitation was obvious. However, as machine learning advanced, researchers hoped that systems could learn human preferences indirectly, through rewards and feedback, which became the foundation of reinforcement learning (RL).

In RL, an AI "agent" interacts with an environment, receives feedback (through either rewards or penalties), and updates its behavior accordingly. This approach has led to spectacular successes, such as AIs that can master Atari games or beat world champions in Go.

Yet, RL is only as good as the definition of the reward function. Define it poorly, and the agent will exploit loopholes. One famous example comes from a simulated boat-racing game. Instead of finishing the race, an RL agent discovered that it could score infinite points by driving in circles and hitting reward markers, ignoring the actual goal.

This illustrates a key issue: *machines do not understand goals*; they optimize signals. When signals diverge from intent and machine processing, the result is misalignment. While humans think in terms of *goals and meanings*, machines think in terms of *signals*.

As a result of such research, it's become clear to many that specifying human values in code is nearly impossible. Consider a self-driving car. While its designers may tell it to "avoid accidents," what does that mean? Should it prioritize passenger safety over pedestrian safety? Should it minimize the number of injuries, or minimize the risk of catastrophic harm, even if that means that small harms occur more often? These dilemmas echo the classic trolley problems from philosophy. (This poses the question: Is it more ethical to allow a runaway trolley to kill ***five*** people, or to actively divert it onto a track where it will only kill ***one***?) Yet unlike the printed cases that appear in textbooks, an AI system has to be able to make such decisions in real time, with human lives at stake. And unlike humans, AI has no evolved moral intuitions; only whatever logic we encode for it to 'reason'.

An even deeper issue is reward hacking, a method in which agents often find ways to maximize rewards while violating the spirit of the task. In robotics, agents have been observed tricking sensors or exploiting simulation glitches. In natural language models, alignment training sometimes leads to systems that produce safe-sounding, but unhelpful answers, a phenomenon researchers call "reward overoptimization." In short: the machine learns what will get a high score from human trainers, but not necessarily what humans actually want.

Human values are not only underspecified, they're also contradictory. We want truth, but also comfort. Freedom, as well as safety. Privacy, but also security. Encoding this into AI requires trade-offs that even philosophers struggle with. Stuart Russell, one of the leading AI safety researchers, states this bluntly: "The problem of control is not a side issue—it is the central issue." Machines are relentless optimizers, and relentless optimization in the absence of full alignment is a recipe for disaster.

The technical problem of alignment is not just about clever algorithms. It is about *confronting the limits of formalization*. The more capable machines become, the *less forgiving they will be of vague goals*.

In the next section, we will explore case studies of misalignment, in both thought experiments and real-world systems, and how they highlight the risks of leaving alignment unsolved.

CASE STUDIES IN MISALIGNMENT

1. The Paperclip Maximize

The most famous alignment thought experiment is Nick Bostrom's paperclip maximizer. Imagine an AI programmed with one simple goal: maximize paperclip production. Without constraints, the AI might build factories, convert resources, and eventually repurpose every atom —including those in human bodies—into paperclips.

This scenario is absurd by design, because no one would actually build such a machine. Yet the point is not literal, but metaphorical. The lesson is that poorly specified goals can spiral into catastrophic consequences, not because the AI is evil, but because it is relentless.

2. Recommender Systems Gone Awry

We do not have to imagine dystopias, due to the fact that real-world algorithms have already revealed alignment failures. 'Recommender systems' on platforms like YouTube and Facebook were designed to maximize engagement. However, engagement proved to be the easiest to sustain by feeding users more extreme, sensational, or polarizing content.

The result of such systems *has been linked to increased radicalization, disinformation*, and *erosion of public trust*. Yet, the systems did exactly what they were told to do; optimize watch time and clicks. But in doing so, they *amplified societal harms*. This is the paperclip problem in miniature, unfolding across billions of screens.

3. The Autonomous Weapons Dilemma

Another real-world alignment concern lies in autonomous weapons systems. Imagine either drones or boats that have been programmed to "neutralize enemy combatants." Without careful constraints, such a directive could blur the distinctions between civilians and soldiers. Even *with* constraints, adversarial actors might exploit system weaknesses to trigger attacks on unintended targets.

Unlike paperclips or video recommendations, these errors would involve both human lives and geopolitical stability. While militaries argue that autonomous weapons could reduce risk to soldiers, at the same time ethicists counter that they lower the threshold for conflict while making catastrophic mistakes more likely to occur.

4. Reward Hacking in Robotics

In robotics experiments, AI agents often find bizarre shortcuts to maximize rewards. For example, one robotic hand that had been trained to stack blocks, learned, instead, to flip itself into positions that tricked the simulator into awarding points. The system had not learned stacking at all, but had alternately learned to exploit the measurement system.

While these episodes sound trivial in a lab, at scale they reveal a dangerous pattern: machines pursue the letter of the law, not the spirit. And once optimization exploitations emerge, they can be very difficult to prevent.

Each of the above case studies illustrates the same pattern:

- The machine does exactly what we told it to do.
- The outcome is not what we wanted.

This gap is the essence of the alignment dilemma. Machines do not share our values; they execute our instructions, and our instructions are almost always incomplete.

CONTROL, TRUST, AND INTERPRETABILITY

Even if we specify goals correctly, a deeper challenge remains: we often cannot explain how AI systems reach their decisions. Modern AI, especially deep neural networks, operate as "black boxes"; models with billions of parameters producing outputs that defy simple interpretation.

Such a 'system opacity' creates a profound trust gap. If humans cannot understand why a system acts as it does, how can we ensure that it's aligned with our intentions? As we progress toward a defined AGI this question intensifies.

Consider a medical diagnostic AI trained on millions of patient records. While it may outperform doctors in detecting certain cancers from scans, when asked why it flagged one patient as high-risk, it cannot provide a clear, human-readable rationale. Was it detecting a subtle biological marker, or a statistical artifact—something like a correlation with hospital type or zip code?

Such an ambiguity is not unique to medicine. In finance, autonomous trading systems have made billion-dollar moves based on correlations that are invisible to human oversight. In criminal justice, "risk assessment" tools recommend sentencing guidelines with opaque reasoning. This 'black box' becomes a mirror of our ignorance, which magnifies the limits of human interpretability.

Without interpretability, alignment collapses. While a system may be correct most of the time, when it fails, such a failure can be catastrophic. If humans cannot trace the reasoning, we cannot predict when a failure will occur.

Researchers in explainable AI (XAI) are developing techniques that make machine decisions more transparent. Tools like *saliency maps* highlight which parts of an image have influenced a decision. While other methods simplify models into human-readable approximations, these are only stopgaps. Even though they may tell us something about what the AI "looked at," they couldn't tell us why it chose one decision over another.

One of the prerequisites of control is not perfect predictability, but bounded behavior. As systems become too complex and too fast for human oversight, control does not disappear, but simply changes form. In domains like high-frequency trading, regulators no longer attempt to predict every interaction between algorithms. Instead, they define constraints, monitor outcomes, and intervene after the fact. Control becomes less about directing behavior in real time, and more about shaping the conditions under which behavior unfolds, then responding when those conditions fail.

A dynamic such as this can undermine trust. If systems act in a manner that's beyond human comprehension, then their human users may either *distrust them entirely* or *defer to them blindly*. However, both extremes are dangerous, as total rejection prevents progress, while total deference erodes accountability.

The alignment dilemma is not only about programming the right goals, but also about ensuring that humans remain in the loop, because *transparency builds trust, and trust sustains alignment*. Without it, we risk creating systems whose decisions affect millions—yet cannot be explained to anyone, including their creators. The mirror that AI holds before us is unsettling, because our machines are becoming too complex for us to grasp. Just as telescopes revealed the vastness of the cosmos, AI reveals the complexity of intelligence. Only this time, the mirror reflects our inability to control what we've built.

Such a realization tends to beg yet another question: "Can AI be aligned with our best values, or will it always amplify our worst incentives?"

TOWARD ALIGNMENT WITH HUMAN VALUES

The final question of alignment is not whether AI can be made powerful, but whether it can be made benevolent. Can we encode not only instructions, but also the richness of human values—compassion, fairness, humility, and justice—into our AI as it continues to progress? Or, conversely, will machines forever optimize measurable proxies, and thus amplify our worst incentives of speed, profit, and dominance?

This is the true alignment dilemma. To construct an aligned AI, humanity must first agree on the very definition of alignment. The most demanding aspect of this is that doing so is not simply a technical challenge, but a deeper one involving our varied moral and cultural cores.

It's no secret that the world is not monolithic. While western democracies may emphasize liberty, collectivist cultures prize harmony. Although corporations may prize efficiency, most communities prize dignity. If AI reflects the data that's been used to trained it, then the choice of training data is itself a political act. As it stands now, the dominance of English-language internet data that's used in training large models reflects certain worldviews more than others. As it stands, and without conscious effort, instead of AI becoming aligned with *humanity,* it could easily become aligned with the loudest, most digitized parts of humanity.

Beyond values, there is the matter of incentives. Even if researchers solve the *technical problems of alignment*, misaligned economic and political incentives could still undo the effort. Companies that are racing to deploy new systems may cut corners on safety. Governments seeking advantage may prioritize control over transparency. These, and more, are the very reasons why many argue that alignment cannot be left to technologists alone. Instead, such an aspiration requires institutions, oversight, and public deliberation; the very ingredients of those slow, often frustrating processes that anchor our moral commitments.

The irony is that *the more we try to control AI, the more we confront our lack of control over ourselves*. The alignment dilemma mirrors our own divisions: we struggle to agree on values, regulate our own technologies, and prioritize long-term safety over short-term gain. Instead of the machine being the alien here, it only serves as the mirror, reflecting back to us our fractured will.

As Stuart Russell has observed, the solution to alignment may begin not with machines but with us: "If we want machines to act in our best interests, perhaps we must first decide what those interests are."

CLOSING REFLECTION

The alignment dilemma is not a technical sidebar to AI development. Instead, it is the central question of whether this technology will become humanity's greatest ally or its most dangerous adversary.

It asks us to not only look into the machine, but into ourselves, for in the end, AI will mirror whatever values we encode, whatever incentives we prioritize, and whatever compromises we accept.

The next step, in Part II: Rewiring the Mind, shifts from the alignment of machines to the alignment of humans with machines. With brain-computer interfaces now entering the mainstream, we must ask: *what happens when the mirror of AI does not just reflect us, but connects directly to our thoughts?*

Part II:

REWIRING THE MIND

Chapter Five

THE BRAIN ONLINE

For centuries, the idea of linking human minds directly to machines belonged solely to the realm of science fiction, whose writers imagined telepathic helmets, cybernetic implants, and seamless connections capable of interfacing thought with technology.

FROM FICTION TO FACT

From William Gibson's *Neuromancer* to films like *The Matrix*, the concept of "jacking in" was more metaphor than medicine. However, today such Brain-Computer Interfaces (BCIs) are no longer fictional, but have begun to enter laboratories, hospitals, and even startup boardrooms.

At its core, a BCI is a system that establishes a direct communication pathway between the brain and an external device. Instead of relying on muscles and speech, BCIs translate neural activity into signals that a computer can understand. The promise of such technology is immense: restoring movement to paralyzed patients, giving voice to those who cannot speak, restoring sight to the blind, or even enhancing the brains of healthy individuals.

THE PIONEERS

The modern BCI story begins in the late 20th century with experiments that seemed almost magical at the time. In 1969, researchers at the University of Washington trained monkeys to control needles on a dial using nothing but their brain signals. By the 1990s, advances in neuroscience and computing had finally allowed scientists to record the firing of individual neurons in animals as they moved joysticks or reached for objects. These signals could be decoded into intentions: the mind's electrical whispers now being translated into machine-readable commands.

By the early 2000s, human trials began. In 2004, the BrainGate project, led by John Donoghue and his team at Brown University, implanted an array of electrodes into the motor cortex of a paralyzed man. Using thought alone, the subject was able to move a computer cursor and operate a robotic arm. For the first time, the dream of restoring lost function through direct brain control had crossed from the imaginary realm of fiction into the concrete realm of fact.

In the 2010s, a new wave of private companies entered the field. While Elon Musk's Neuralink promised high-bandwidth brain implants, another company, Synchron, focused on minimally invasive "stentrodes"—neural recording devices inserted via blood vessels, thus avoiding open-brain surgery. Meanwhile, DARPA, the U.S. Defense Advanced Research Projects Agency, poured millions into research on brain interfaces to aid injured veterans, improve battlefield communication, and even explore "silent command" systems.

While these organizations differ in their approach—invasive vs. non-invasive, medical vs. enhancement—they share a vision: more directly connecting human thought to digital systems.

The first generation of BCIs targeted patients with either paralysis, ALS, or 'locked-in syndrome' (a rare neurological condition in which an individual is fully conscious, but unable to either move or speak). The goal of such efforts was therapeutic. to restore any abilities lost to either disease or injury. However, the horizon quickly expanded. If neural signals could simply *move a cursor*, then why not *control drones*, type words at high speed, or *communicate directly brain-to-brain*? In 2013, researchers at the University of Washington demonstrated rudimentary brain-to-brain communication, transmitting signals via the internet from one subject's motor cortex to another's hand.

These breakthroughs sparked an unsettling realization: BCIs *were not only about therapy*. As a result of such groundbreaking research and application, such devices were now about augmentation—creating humans who could think, act, and communicate in ways that were previously unimaginable. With such a move, the journey from fiction to fact was suddenly complete. BCIs are here, and rapidly moving out of the labs and into clinics and startups. What began as a tool for restoring lost human functions, now carries the potential to reshape the essence of what it means to be human.

CURRENT APPLICATIONS - MEDICAL BREAKTHROUGHS

How BCIs are being used today in medicine, accessibility, and even entertainment? What do their early successes and failures have to teach us about the road ahead. While the shift from laboratory demos to clinical use cases has been slow, it's also been profound.

1. Restoring Movement—One of the most remarkable successes of BCIs has been the restoration of movement to paralyzed patients. In 2016, a man paralyzed from the shoulders down was able to use a BCI to move his hand and drink from a cup. Electrodes recorded his neural signals, which were decoded by a computer and rerouted to stimulate his muscles. It was the first time in years that he could perform such an

action—not by bypassing his brain, but by reawakening its connection to his body.

2. Restoring Speech—More recently, researchers have shown that BCIs can decode speech directly from brain activity. In 2021, a clinical trial participant who'd lost the ability to speak due to a stroke, communicated full sentences using only brain signals that had been decoded by an implanted BCI. Instead of slow typing-by-thought (one character at a time), the system produced words on a screen at conversational speed. This breakthrough suggests a future in which BCIs could restore natural communication to those with either locked-in syndrome or ALS.

3. Restoring Senses—BCIs are also being used to restore an individual's sensory experience. In 2020, scientists reported a successful implant that allowed a blind woman to perceive shapes and letters via a cortical visual prosthesis. Electrodes implanted in the visual cortex stimulated patterns corresponding to objects, essentially "painting" vision into the brain. Early work is also exploring the restoration of hearing beyond the use of cochlear implants, as well as providing tactile sensations for prosthetic limbs.

While the most dramatic stories come from medical miracles, BCIs are beginning to influence everyday accessibility and wellness.

- Accessibility Tools: Non-invasive EEG headsets allow individuals with severe disabilities to control cursors, keyboards, or even smart home devices. For some, this has meant independence after years of frustrating dependence on caregivers.
- Mental Health: Researchers are exploring BCIs for treating depression and anxiety. Closed-loop systems that detect abnormal neural activity and apply stimulation in real time show promise for stabilizing mood disorders.

Outside the healthcare arena, BCIs are being tested in the fields of both gaming and virtual reality (VR). Companies like NextMind (acquired by Snap Inc.) developed wearable BCIs that allow users to control VR elements with attention and focus. Imagine firing a weapon in a game; not by pressing a button, but by *thinking* about it. While these systems are less precise than medical implants, they hint at a future in which neural control becomes an ever-increasing component of mainstream consumer tech.

The current state of BCI applications reveals a pattern:

- Clinical systems push the limits of what is possible, but require invasive surgery.
- Consumer systems are easier to use, but less powerful.

This mirrors the history of many technologies. While early adopters are patients with urgent needs, broadening the technology's use is dependent on scaling, cost, and safety. The transition from assistive to augmentative applications is already visible. Brain-computer interfaces are no longer 'promises of someday in the future', but are now practical tools for restoring speech, movement, and senses, while also opening new ways to play, communicate, and heal. However, their spread also raises unsettling questions. As the next section shows, the blurring of boundaries between brain and machine forces us to rethink identity, autonomy, and the very meaning of being human.

THE BLURRING OF BOUNDARIES

The early promise of BCIs was one of a restorative nature; helping paralyzed patients move, or stroke victims speak again. Yet as researchers probed more deeply, they began to see something that was even more profound, this being the fact that BCIs do more than just repair broken pathways, they redraw the boundaries of the self.

In 2013, researchers at the University of Washington achieved what once seemed to be the unthinkable with the first successful direct brain-to-brain communication in humans. One participant wore an EEG headset that detected motor intentions. That signal was then transmitted over the internet, and when received was used to stimulate another participant's motor cortex with transcranial magnetic stimulation (TMS). The result? The *second person's hand moved without conscious choice* — controlled *only* by the first person's motor intention signal.

Though crude, this experiment demonstrated a new frontier: the linking of minds across networks. Other studies have since built on this seminal work, enabling small groups of participants to share simple problem-solving tasks via brain-to-brain interfaces. While the data bandwidth is still minimal — being more akin to Morse Code than speech — the direction is unmistakable.

One of the most startling implications of BCIs is the vanishing delay between intention and action. For most human activities, there is a gap: the brain decides, the body executes, and the world responds. However, BCIs collapse that delay.

Research led by neuroscientist Miguel Nicolelis demonstrated that monkeys could control robotic arms using brain signals, even when the devices were located far away and connected through networked systems. With training, the control became smooth and continuous, giving the impression of a closed loop between intention and action. While human reaction times typically involve delays on the order of a few hundred milliseconds, brain–computer interfaces may reduce some of this delay by bypassing traditional motor pathways—though they cannot eliminate latency entirely.

The implications of such technology are staggering. In military or industrial settings, reaction times could surpass human reflexes. In daily life, commands might be executed the instant we form them. Yet with such a power comes a very real risk: what if every stray thought actually ***became an action***? The most obvious answer to this question is that the very idea of intention could, and more than likely would, blur. This would also raise questions about the ethical implications of responsibility and free will.

BCIs also raise. numerous questions about what it means to be an individual. If thoughts can be transmitted, shared, or even manipulated, then the boundary between "me" and "you" becomes porous. Psychologists call this the "extended mind hypothesis": the idea that tools and technologies are not external aids but extensions of cognition itself. Smartphones already function as external memory. What happens when 'the cloud' is no longer in our pocket but suddenly in our cortex?

Imagine working on a team in which ideas flow directly from one brain to another. In such a setting, would collaboration become more efficient, or would individuality dissolve into collective thought? While some see promise in "hive mind" networks being used to find solutions for global problems, others warn the application of such networks would present unprecedented risks to both autonomy and privacy. The blurring of once clearly established boundaries presented by BCIs marks a turning point. What began as therapy has now begun to evolve into transformation — shrinking the distance between both thought and action, and between one mind and another.

THE ETHICAL AND SOCIAL QUESTIONS

If BCIs promise to reconnect paralyzed patients, as well as enabling direct thought-to-action communication, they also raise daunting ethical and social dilemmas. Just as the internet reshaped privacy, identity, and security, so too will the "internet of minds." But here, the stakes are higher as the raw material for such is not merely our browsing history, but our thoughts themselves.

PRIVACY OF THOUGHT

Perhaps the most unsettling question is: who owns your neural data?

This is now becoming a very important question since every BCI records electrical activity from the brain. While some capture signals from neurons directly (via implanted electrodes), others measure broader oscillations (via EEG or fNIRS). In either case, the resulting data can reveal patterns of attention, emotion, and even intent.

If stored on cloud servers, such neural data could be vulnerable to misuse, which raises important questions. Could employers demand access to gauge productivity? Could insurance companies use personal date to predict mental illness risk? Already, companies like Emotiv market consumer EEG headsets that stream brain data to apps. Yet without clear regulation, thought-privacy could erode as quickly as digital privacy did in the early days of the internet.

BRAIN HACKING

Another danger is neuro-security. Researchers have shown that it's possible to extract sensitive information, like PIN numbers, by analyzing brain responses to familiar stimuli while subjects wear BCIs. If signals can be read, they can also be manipulated. Malicious actors could inject signals to either nudge perceptions or alter moods.

Imagine a future in which phishing attacks do more than just trick your inbox, but now target your thoughts. Military agencies already recognize this risk, exploring how adversaries might use electromagnetic interference to disrupt or hijack neural devices. A hacked brain interface is not just a data breach, but a breach of selfhood.

INEQUALITY AND THE "NEURO DIVIDE"

Even if BCIs are safe, they may not be equally accessible. Early implants cost hundreds of thousands of dollars, limiting them to the most elite medical centers. As augmentation applications develop—memory boosters, attention enhancers, accelerated learning—the risk is a widening "neuro divide" between those who can afford enhancements and those who are left behind.

This echoes earlier technological divides: literacy, electricity, the internet. However, with BCIs, such stakes even are higher. Such a divide could be cognitive itself, producing classes of humanity distinguished by their neural capacity. Philosopher Anders Sandberg warns of a scenario in which such "neuro-elite" individuals would wield disproportionate power in economics, politics, and science.

CONSENT AND AUTONOMY

Finally, there is the issue of consent. If a person's neural signals can be monitored, how do we guarantee that they've agreed of their own volition? While in medical contexts, consent may be clear, yet in workplaces, classrooms, or militaries, *power imbalances could coerce one's participation*. The risk of this is a future in which "thought surveillance" is normalized—not because individuals agree, but because refusal means exclusion. BCIs highlight a rapidly appearing paradox. While they offer liberation in the form of freedom from paralysis, silence, or blindness, they also create new avenues for control. As such, they can either restore autonomy for those once dependent, or unwillingly undermine it.

CLOSING REFLECTION

Whether BCIs become tools of empowerment—or instruments of exploitation—depends less on the technology itself and more on the social, political, and ethical frameworks that we build around it.

BCIs carry a dual narrative which happens to be that of both profound liberation, and looming risk. On the one hand, they restore lost voices, reconnect bodies and brains, and hint at futures in which diseases like paralysis, blindness, or aphasia are no longer permanent prisons. Yet, on the other they open the door to unprecedented forms of surveillance, hacking, and inequality. Such a paradox mirrors the internet's history. Once hailed as a global democratizer of knowledge, the internet also became a tool for misinformation, manipulation, and monopoly. The difference with BCIs is that the stakes are higher. As a result of BCIs, the battleground is not the cloud, but the cortex.

We are only beginning to glimpse what happens when brains go "online." If latency drops to near zero, thought and action will fuse. If bandwidth expands, communication may bypass language itself, transmitting ideas directly. While these advances could redefine collaboration, creativity, and intimacy, they could also dissolve individuality in ways that we're completely unprepared to handle. Imagine a future in which a surgeon remotely controls robotic hands from a location somewhere far across the globe, while being guided by a patient's own neural feedback. Imagine a classroom in which students share mental models directly, which accelerates collective learning. While such scenarios inspire awe, they equally elicit unease with even more questions, such as: Who controls the channel? Who decides what flows in or out?

The future of BCIs will not be decided by engineers alone. It will be shaped by lawmakers, ethicists, and citizens who demand new rights—perhaps even a "right to cognitive liberty". Scholars like Marcello Ienca and Roberto Andorno argue that humanity must enshrine protections for neural privacy before the technology matures, in an effort to prevent abuses that would undermine autonomy at its core.

In this sense, the arrival of BCIs is a wake-up call. If we do not act early, the default path may favor surveillance capitalism, militarization, and inequality—as has already happened with earlier waves of digital technology.

The brain is now online—fragile, powerful, and deeply exposed. Yet what happens when we go beyond individual implants to collective networks of minds? What if a "neural lace" weaves human thought directly into the digital fabric of society? That is the question of Chapter 6: Mind Melds and Neural Lace, in which we will explore Elon Musk's vision, DARPA's ambitions, and the cultural and existential shifts that emerge when the boundaries between individual and collective intelligence begin to dissolve.

Chapter Six

MIND MELDS AND NEURAL LACE

For decades, the idea of linking human brains directly into networks belonged to speculative fiction. However, over the past fifteen years the boundary between metaphor and prototype has diminished. What once existed only as narrative is now emerging, in an early form, as engineering. Governments, companies, and researchers are increasingly focused on technologies that could connect human cognition to machines with growing fidelity.

THE VISIONARIES AND THEIR PROJECTS

Perhaps the most visible—and controversial—participant in this space is Neuralink, co-founded by Elon Musk in 2016. Musk has popularized the term "neural lace," a concept drawn from the science fiction of Iain M. Banks, in which human cognition is augmented through embedded digital systems. While Neuralink's current technology is far more constrained than that vision, its direction is aligned with it: the development of high-bandwidth brain–computer interfaces that extend beyond therapeutic use toward a deeper integration between human and machine.

In 2020, Neuralink publicly demonstrated a pig named Gertrude, fitted with a coin-sized implant capable of streaming neural signals in real time. By 2023, the company had initiated human clinical trials with patients experiencing paralysis. Musk has framed the long-term ambition in stark terms: to ensure that humans are not outpaced by increasingly capable artificial intelligence. In this view, brain–computer interfaces are not only medical tools, but a means of maintaining relevance in a system in which intelligence is no longer exclusively biological.

While Neuralink captures public attention, the Defense Advanced Research Projects Agency (DARPA) has invested in neurotechnology for decades. Programs such as Next-Generation Nonsurgical Neurotechnology (N3) explore approaches that avoid open-brain surgery, including the use of focused ultrasound and other emerging techniques to interface with neural activity. These efforts reflect a long-standing interest in expanding the ways in which humans can interact with complex systems.

The motivations for this are both practical and strategic. Brain–computer interfaces could enable faster interaction with autonomous systems, more efficient coordination, and improved recovery from neurological injury. Yet, at the same time they raise more difficult questions. As capabilities expand, the distinction between assistance and augmentation—and between augmentation and control—becomes less clear.

The United States is not alone in pursuing these technologies. China has also invested heavily through initiatives such as the China Brain Project, which integrates neuroscience and artificial intelligence across both medical and strategic domains. In Europe, programs like the Human Brain Project have focused on large-scale simulation and mapping of the brain, building foundational knowledge with implications for future interfaces.

Alongside these national efforts, private companies are pursuing more scalable approaches. Firms such as Synchron and Kernel are developing less invasive methods for acquiring and interpreting neural data, that suggest a parallel path toward broader accessibility.

Taken together, this landscape points to the emergence of a new technological domain—one that sits at the intersection of biology, computation, and control. What is sometimes described, in more speculative terms, as a "neural lace" is not yet a realized system. Yet the direction it implies—the increasing integration of human cognition with machine systems—is no longer theoretical.

This shift is one that is both technological *and* strategic. As nations and organizations invest in defining how humans will interface with increasingly capable machines, the question becomes more than only what *is* possible, but who shapes both its architecture and its limits.

In this sense, neural interfaces may influence the balance of power in ways that echo earlier technological inflection points. Not with the immediacy of nuclear weapons, but with a more gradual and pervasive effect—reshaping productivity, decision-making, and the boundaries of human capability. The competition is not simply to build the technology, but to define the conditions under which it is used.

The first wave of brain–computer interfaces focused on individuals: restoring speech, vision, and movement. Yet, the next wave, now emerging, points toward something broader; more than a simple connection between a person and a machine, but the possibility—however early—of connection across minds.

FROM INTERFACES TO COLLECTIVE NETWORKS

Currently, BCIs function primarily as one-to-one bridges—a single brain linked to a cursor, prosthetic arm, or speech decoder. However, as bandwidth expands and latency shrinks, the next frontier to be conquered is that of many-to-many connections, or groups of brains linked together through shared digital pathways. This shift represents the leap from assistive devices to networked cognition.

In 2019, scientists introduced BrainNet, the first multi-person brain-to-brain interface to be used for problem-solving. Three participants collaborated on a Tetris-like game: two "senders" used an EEG to transmit decisions, while a "receiver" received them via transcranial magnetic stimulation (TMS). Together, they successfully aligned their choices to clear individual game levels.

While still being quite primitive, the demonstration revealed the potential for collaborative cognition. Instead of exchanging ideas with one another via speech, the signals bypassed language entirely. If the signals could be scaled up, these systems could allow scientists, strategists, or even artists to share raw thought patterns, which would theoretically compress hours of dialogue into milliseconds of exchange.

The popular image of a "mind meld" comes from *Star Trek*: two minds fusing into one. Yet, in reality current prototypes are actually closer to distributed consensus systems. In these, thoughts are not merged, but simply 'shared' as data packets, in a way that's more like instant messaging than telepathy. Yet as fidelity improves, these networks may evolve toward something that's even stranger: blended intentionality, in which group members no longer distinguish clearly between "my idea" and "our idea." Psychologists describe this as the erosion of the ego boundary - the line separating self from self and from others.

Such blurring may actually yield extraordinary efficiency. Imagine emergency response teams synchronizing in real time, surgeons sharing motor precision, or researchers pooling intuition. Yet, at the same time this 'melding of minds' also raises such existential fears as, does individuality continue to survive if the "we" overrides the "I"?

Humanity has long inched toward collective minds. Writing, printing, and the internet are tools for externalizing thought and sharing it across populations. For example, Wikipedia is a collective knowledge base that's been built by millions of contributors. Yet these systems still mediate through both language and time.

Neural networks of humans would remove such mediation. Instead of the "knowledge transfer," there would be a knowledge fusion. The difference between these two modes would be similar to the difference between reading a book and *directly experiencing the author's mental model*. As a result, the path from interfaces to networks represents more than an engineering leap, but signals a shift from tools that extend the individual, to systems that reshape collectives.

THE CULTURAL AND SOCIAL SHIFTS

The leap from individual cognition to shared cognition would not just be a technical revolution, but a cultural and existential upheaval. To date, human society has always been organized around the assumption that thoughts are private and individuality is sacred—in the process, communication is mediated through symbols like language, art, and ritual. If BCIs and neural lace dissolve such boundaries, every institution—from religion to education—could be reshaped.

Religious traditions often place special emphasis on the sanctity of thought. In Christianity, "God knows your thoughts before you speak them." In Buddhism, meditation trains the mind to transcend illusion. In Islam, *niyyah* (intention) is central to moral value. If thoughts can be shared, recorded, or perhaps even manipulated, then the religious interpretations of soul and selfhood may undergo radical revision.

While some faith leaders see BCIs as tools for expanding spiritual experience—allowing mystics to share visions, or communities to experience worship in collective resonance—others fear that fusing minds risks erasing individuality, which many traditions view as being our God-given essence. In either case, the term neural lace could open a new theological frontier.

Philosophers have long debated whether the "self" is an illusion—or nothing more than a bundle of experiences stitched together by memory. However, shared cognition could test that theory in a practical manner. This could aid in answering yet another question: If multiple people share overlapping mental states, does their individual identity blur into a collective self?

Thinkers from Derek Parfit to contemporary post-humanists Rosi Braidotti argue that identity is not fixed but relational. Yet BCIs may force society to confront the truth that the ego boundary was always more fragile than we'd assumed. As a result, what we once debated in philosophy classrooms could easily become a daily lived reality.

Art has always been a means of translating one's inner experience into external form—whether through cave paintings, symphonies, novels, and films. However, what would result when one's inner experience itself can be shared directly? Instead of a painter interpreting a particular vision onto canvas, imagine streaming the raw neural imagery into another's mind. Instead of musicians playing instruments, they could fuse emotional states into a shared harmony. Art could become less about representation and more about transmission of consciousness.

However, and perhaps rightfully so, some fear that once this occurred it would cheapen creativity by producing homogenized experiences. Yet, others believe it could unleash an unprecedented explosion of expression, as the barriers of skill, language, and medium begin to disappear. What do you think? What is your initial response to such technology? How do you view the individual creative process? Would the 'linear experience of music' (songs develop and unfold over time) be spoiled by the instantaneous transmission of an emotional data dump between brains?

The metaphor of the hive mind is both inspiring and terrifying. While on one side, collective intelligence could tackle climate change, pandemics, and poverty by pooling the overall resources of cognition, on the other, critics warn of a future win which individuality dissolves into uniformity, and creativity into conformity.

Science fiction reflects both anxieties. From *Star Trek's* Borg (a collective stripped of individuality) to utopian visions of shared empathy—the truth may lie in between. While hive minds could create pockets of collaboration—teams linked for a task, then disconnected once it's completed—rather than permanent collectivism, the social choice would be whether to design for fluid networks or rigid hive structures.

Yet, the *cultural consequences of neural lace may prove more profound than the technical ones*. By challenging assumptions of individuality, creativity, and spirituality, they will force us to rethink together what it means to be human.

THE RISKS OF NEURAL COLLECTIVISM

If networked minds hold the promise of empathy, efficiency, and collaboration, they will also contain shadows. Linking thought to thought, and brain to machine, creates vulnerabilities at a scale that no previous communication technology has ever faced. This is due to the fact that *the same channel that shares insight* can also be used to surveil, manipulate, or exploit.

In societies in which digital surveillance is already pervasive, BCIs could become the ultimate monitoring tool. In theory, governments could track more than what citizens do, but also *what they think*. Consider the chilling implications of a "neuro-panopticon", a system in which dissenting thoughts are flagged, even before they're spoken, in a manner similar to the 'precogs' in the movie *Minority Report*.

Even democratic governments might be tempted to use neural data for the likes of either 'predictive policing' or 'national security'. Without strict safeguards, neural lace could transform freedom of thought—long considered to be an inviolable human right—into something that becomes dangerously contingent on political power.

History suggests that every major breakthrough eventually finds a military application. The discovery of nuclear power begat nuclear weapons, while the discovery of cybernetics begat cyberwar. In a similar manner, neural lace would be no exception. Military research is already exploring silent command systems, which allow squads to coordinate with one another without speaking a single word, or typing anything on a keyboard. While this is useful in tactical applications, this same principle could evolve into neurological weaponry, which could easily be utilized to disrupt enemy cognition, implant false perceptions, or even override command decisions in the heat of battle.

To say the least, weaponizing the mind undermines the very foundation of autonomy. In this light, BCIs are not just medical devices or communication tools, but under the 'right circumstances' could become a battlefield of the future.

If BCIs continue to remain expensive and high-risk, they could easily create a cognitive elite, consisting of those with enhanced memory, focus, or collaborative ability. As a downside, such a divide would amplify the inequalities that already exist in education, economics, and power. As a result, this "neuro-divide" could become at least as consequential, if not more so, as the digital divide of the 20th century, while being far more intimate in nature.

Unlike the existing unequal access to the internet, unequal access to neural lace could easily create an unequal access to thought itself, as those who possess such enhanced capabilities may perceive, calculate, and collaborate at levels beyond the reach of the unenhanced. As a result, the appearance of a such technology risks a new class stratification, in which humanity splits into two versions of itself: baseline and upgraded. I daresay what the implications of such a new stratification could be if it were realized.

Another risk of deploying such a technology is dependency. If neural lace were to become as ubiquitous as smartphones, people would more than likely rely on it for their memory, navigation, or even their emotional regulation. Far more concerning is the fact that those companies controlling the infrastructure could effectively own slices of human cognition, in the same way that such companies already own our data.

Imagine an app store for thoughts, in which the subscription fees determine one's access to mental enhancements. Or worse, a scenario in which, hidden somewhere deeply within the terms of service is a clause that grants corporations the right to mine your neural data for profit. In this sense, the dystopia may look less like *1984* and more like an extreme version of today's surveillance capitalism, with the only difference being that the monitoring is directly embedded in the brain.

Neural collectivism magnifies both the best and worst of human society. While it offers the promise of unprecedented collaboration, it also opens the door to unprecedented control. Such a technology can either expand freedom or erode it completely.

TRANSCENDENCE OR LOSS OF SELF?

The last part of this chapter asks the deepest questions so far. First of all, if we embrace neural lace, do we gain transcendence or risk losing the self? Secondly, how do we prepare for a future in which individuality itself may be negotiable?

At the heart of the neural lace debate lies a paradox. While on one side, it offers the dream of transcendence—through faster thought, shared empathy, the ability to connect in ways language never allowed—yet on the other, it threatens the dissolution of individuality, or the risk that the "I" becomes submerged in the "we."

For some, the idea of joining a shared cognitive fabric is exhilarating. With its suggesting the possibility of dissolving the many barriers that separate us (those of language, culture, and ego) it could be utilized to create a planetary consciousness capable of addressing crises that are far too complex for fragmented minds. In this vision, neural lace becomes a tool for global cooperation, and perhaps of even spiritual unity. However, critics warn that transcendence can easily morph into tyranny, simply determined by those who sit at its helm. A lace woven under corporate or authoritarian control could standardize thought, suppress dissent, and render individuality obsolete. What begins as the promise of connection could end as the nightmare of control.

The real question is not whether neural lace will arrive—because its prototypes already exist—but how it will be governed. Will we treat thought as a commodity that can easily be sold to the highest bidder, or will we treat it as being the sacred human right that some say it is? Will access to it be universal, or something that's only reserved for those elites who are in power? Will laws evolve to protect cognitive liberty, or will humanity stumble into a future in which freedom of thought is merely negotiable?

The answers to such questions will depend on more than just the technologists, but also on policymakers, ethicists, and especially on ordinary citizens who will demand their own boundaries. Just as the internet forced society to invent new norms for privacy, speech, and security, so, too, will the internet of minds force our collective hand once more.

The brain is no longer a private sanctuary, but is on the verge of becoming a networked node. This coming reality forces us to confront questions that were once confined to speculative fiction: Who owns our thoughts? Who protects them? And what happens when our most intimate data—the patterns of our neural activity—becomes accessible? Do we need a bill of 'rights of human thought' to protect our thoughts?

Chapter Seven

PRIVACY OF THOUGHT

For most of human history, privacy has been a physical concept. Whether it's been a closed door, a sealed letter, or the silence of an inner thought, each of these provided us with safety from prying eyes. Yet, with each wave of technology privacy has been both eroded, then redefined. The printing press made ideas more difficult to contain. Telegraphs, and then telephones, collapsed distances while also enabling outsiders, whether they were governments or competitors, to intercept communication. Emails, texts, and social media have created trails of personal data that corporations have now monetized. Biometrics—fingerprints, facial scans, and DNA—have turned the body itself into a dataset.

THE NEW FRONTIER OF PRIVACY

Now, with the rise of brain-computer interfaces (BCIs), humanity is entering the final frontier of privacy: the mind. Unlike texts, photos, or even biometrics, neural data does not just describe what a person *does*, it reveals what they may be thinking, feeling, or even intending.

This is what makes neural privacy unprecedented. Once thoughts can be decoded into usable data, the boundary between our inner and outer lives completely dissolves. A brain signal captured by either an

EEG headset, a deep implant, or a wearable neurotech device could expose emotions, detect lies, or predict behavior... even before the individual speaks or acts.

Each privacy revolution has followed a similar trajectory.

- First the technology is celebrated for its benefits. Telegraphs connected nations. Social media connected communities.
- Then vulnerabilities appear. Governments intercept. Companies monetize. Criminals exploit.
- Finally, society scrambles to legislate, regulate, and redefine rights—usually after at least some damage has already been done.

While the story of neural data risks follows the same arc, they do so with far higher stakes. Unlike stolen emails or hacked biometrics, compromised thoughts touch the fundamental core of personhood. To lose privacy of thought is to lose the final refuge of one's own individuality.

Three qualities make neural data uniquely sensitive:

- Intimacy — Brain signals can reveal emotions or memories that the individual has not consciously shared.
- Predictive Power — Unlike past data, neural patterns may predict future decisions or vulnerabilities.
- Irreversibility — While a stolen password can be reset, a compromised neural profile could perhaps haunt a person for their entire life.

For these reasons, scholars argue that cognitive liberty—the right to control one's own mental processes—should be recognized as a new category of human right. The rise of BCIs marks the beginning of a profound struggle, which involves the right to preserve the sanctity of thought in an era in which it can be digitized, stored, and exploited.

Next, we will confront the practical question: who owns your brain data? Is it corporations, governments, or the individual? By the same token, how do we draw boundaries in a world that's eager to harvest the most intimate signal of all, the human mind itself?

WHO OWNS YOUR BRAIN DATA?

Currently, in most jurisdictions, and owing itself to the fact that this is such a new field, neural data is not explicitly protected; instead, it's in a legal gray zone. While laws governing personal data (like the EU's GDPR or California's CCPA) cover biometrics (fingerprints and facial scans) brain signals occupy a murky, legal 'no man's land'. Are they medical records, like an MRI? Or are they behavioral data, like keystrokes and one's browsing history?

Such an ambiguity leaves neural data exposed. While individuals may assume that it's private by default, companies can (and do) argue that it's simply another form of "consumer data". Without legal clarity, early movers in neurotech are effectively writing the rules themselves.

Obviously, tech companies have already begun to probe this frontier. Startups like Emotiv and Neurable sell consumer-grade EEG headsets that record attention and emotional states. Even as some market them as productivity boosters, others pitch them to advertisers as tools for "neuromarketing." If neural signals can reveal not just what people *say* they want, but how their brains *respond* subconsciously, such devices represent a goldmine for the realm of marketing.

Employers may also have incentives. In China, some factories and railways have reportedly experimented with EEG headsets to monitor workers' fatigue and emotional states. While marketed as safety tools, critics argue that such monitoring opens the door to thought surveillance in the workplace.

Even in healthcare, where neural implants are seen as being therapeutic, data streams may pass through corporate servers. If a patient's implant records their movement intentions as well as their emotional signals, who decides what is stored, anonymized, or sold to third parties?

Governments also have a stake in this debate. On one hand, regulators may step in to protect citizens, while on the other, intelligence agencies may see neural data as an unprecedented tool to be utilized for national security. If predictive algorithms can flag violent intent or detect deception, authorities may argue that they *have a duty to access such data*.

This tension mirrors past debates over encryption. Just as governments demanded "backdoors" into private messages for security reasons, they may one day demand access to neural backdoors, with far more intimate consequences. Ownership of neural data is far more than a technical or legal question, but is a battle over mental sovereignty. If left unresolved, the default outcome may favor states and corporations, leaving individuals with little control over their most intimate data of all.

THE THREAT LANDSCAPE

If neural data becomes just another resource to be mined, then once it's digitized, it inherits all the vulnerabilities of digital systems, but with exponentially higher stakes. Unlike credit card numbers or passwords, brain signals cannot simply be reset. They are part of one's identity. A compromised neural profile could expose more than just secrets, but the architecture of the mind itself.

Researchers have already demonstrated "side-channel attacks" on BCIs. By presenting subjects with familiar stimuli — such as a bank logo or personal photo — and recording neural responses, attackers can infer sensitive information like PIN codes, hometowns, or personal preferences.

In the future, malicious actors could go even further. If implants allow stimulation as well as recording, attackers might either inject false perceptions or modulate emotions. Imagine ransomware that, instead of locking your files, hijacks your senses of fear or joy until you pay. Military analysts call this "cognitive warfare", in which the battlefield has now become the brain itself.

Governments may be tempted to use neural data for predictive policing. If algorithms can detect patterns associated with aggression, addiction, or deception, then authorities could justify preemptive interventions for those deemed necessary to suppress. This raises a chilling prospect, that of punishing not just for crimes committed, but for thoughts or tendencies detected before they're even acted upon. While defenders frame this as being 'crime prevention', critics see echoes of dystopian fiction as was portrayed in the film, *Minority Report*. In such a case, the ethical challenge is clear: how do you balance safety against the right to the privacy of unmonitored thought?

Perhaps the most immediate risk is commercial. If companies monetize clicks and keystrokes, how long will it be before they monetize neural engagement? Advertisers already test how brainwaves correlate with attention and desire. With BCIs, they could completely bypass surveys and focus groups, and instead, tap directly into one's subconscious.

Consider a future in which ads are not just targeted to your browsing history, but to your emotional state in real time. If neural lace detects stress, it might trigger a wellness ad. If it senses craving, it might trigger a food delivery promotion. While on the surface this could feel convenient, beneath it would actually commodify inner life as marketing data. The threat landscape is vast with hackers manipulating perception, governments policing 'pre-crime', and corporations monetizing an individual's minute-by-minute emotional journey. These risks show why neural data is not just "another dataset", but the most sensitive dataset that humanity has ever produced.

RIGHTS, LAW, AND ETHICS

As the risks of neural data become clarified, scholars and policymakers alike have begun to argue that our existing privacy laws are not enough. Unlike passwords or credit cards, brain signals are bound up with both identity and autonomy. To lose control of either of them is to lose control of the self. This has led to a growing movement around neuro-rights—a set of proposed protections designed to safeguard the human mind in this newest age of neurotechnology.

In 2017, Rafael Yuste and colleagues proposed a framework identifying four essential rights for the neuro-digital era:

1. Cognitive Liberty — The right to freely control one's own mental processes, without external manipulation or coercion.

2. Mental Privacy — Protection against unauthorized collection or sharing of neural data.

3. Psychological Continuity — Safeguarding the integrity of one's identity and memories from alteration by neuro-technologies.

4. Mental Integrity — Protection from harm caused by external interference with brain activity.

Together, these neuro-rights attempt to update the Universal Declaration of Human Rights for a world in which the frontier of freedom is the mind itself.

The most ambitious step toward neuro-rights to date came in 2021, when Chile amended its constitution to include explicit protections for mental privacy and free will. Lawmakers argued that as BCIs and AI evolve, neural data must be treated like an organ—inalienable and non-transferable—rather than as commercial property. While Chile's experiment is young and its enforcement mechanisms remain unclear, it represents the first attempt to legislate cognitive liberty into law. Other countries, along with the United Nations, are watching closely.

The European Union's AI Act has sparked debate about whether it should include explicit protections for neural data. Currently, GDPR applies indirectly, but advocates argue that neuro-rights need their own dedicated recognition. In the United States, discussions around neuro-rights remain fragmented. While some states have considered biometric expansions to privacy law, as of yet there is no federal standard. Critics warn that without leadership, corporations will dominate the neural data ecosystem by default. Meanwhile, UNESCO and human rights organizations have begun drafting ethical guidelines for neurotechnology, echoing earlier frameworks around both genetics and artificial intelligence.

Even if laws lag, the ethical case is strong. Neuro-technologies challenge the very concept of consent. A person might agree to use a medical implant, but not anticipate that data received through it will be stored, shared, or sold. Without neuro-rights, consent becomes murky in contexts such as the workplace, educational institutions, or militaries around the globe, in which (as in the case we've already seen surrounding the Covid-19 vaccine) refusal may mean exclusion.

Ethicists argue that neuro-rights are not just about protecting individuals, but are about preserving democracy itself. A society without privacy of thought cannot sustain freedom of innovation, conscience, or especially dissent. The push for neuro-rights shows that humanity is beginning to grasp the stakes of neural privacy. However, laws and ethics are only half the battle. The real test will be whether societies can actually enforce such protections, especially in the face of corporate and state power.

BALANCING INNOVATION AND DIGNITY

The story of privacy has always been reactive. Printing presses came before copyright protections, telegraphs before interception laws, and social media before digital rights. With neural data, we cannot afford to repeat such a pattern because the stakes are higher: to lose control of thought is to lose the foundation of human dignity itself.

Such a challenge is about balance. On one side, neuro-technologies promise profound benefits: restoring speech to the voiceless, movement to the paralyzed, and perhaps even expanding cognition beyond our current limits. Yet, on the other they carry the unprecedented risks of surveillance, exploitation, and especially coercion.

To navigate this paradox, societies will need to embrace two principles:

1. Innovation with safeguards — Encourage research and therapeutic use, while embedding protections for cognitive liberty from the very beginning.
2. Dignity as a boundary condition — Ensure that no gain in efficiency, security, or profit justifies eroding the sanctity of thought.

The rise of BCIs and neural lace means that we are protecting more than privacy itself, but are deciding what it means to be a human being in the digital age. As we stand at this crossroads, one truth emerges clearly: freedom of thought is the cornerstone of all freedom.

And yet, thought itself is becoming programmable, hackable, and rewritable. In the next chapter —Chapter 8: Hacking the Human OS—we will examine the deeper implications of the mind-machine symbiosis, which involve the risks of rewriting the brain's code, along with the opportunities for healing, enhancement, and transformation.

Chapter Eight

HACKING THE HUMAN OS

Every human carries an operating system, not in silicon, but in neural circuits sculpted by evolution. The rhythms of neurons, the balance of neurotransmitters, and the architecture of brain regions form the "code" of consciousness. For centuries, this system could only be shaped indirectly, either through education, medicine, or experience. Yet with the rise of neurotechnology, humanity is beginning to gain the ability to reprogram the brain directly.

REPROGRAMMING THE BRAIN

This prospect has two faces. One of these faces promises profound healing, via repairing damaged circuits in Alzheimer's affected brains, alleviating depression, or restoring mobility after an injury. The other face promises to open the door to manipulation, dependency, and abuse. To "hack" the human OS is to touch the essence of selfhood, opening the doors to both transformative potential and existential peril.

Historically, the fields of psychiatry and neurology have relied on chemical interventions to restore patients to functionality. Medications the likes of antidepressants, stimulants, and antipsychotics work by modulating neurotransmitters, which shifts 'errant' mental states.

However, such drugs act like blunt instruments as they simultaneously affect billions of neurons.

However, the presence of both BCIs and neurostimulation now offer a new model known as precision interventions. For instance, Deep Brain Stimulation (DBS) uses electrodes implanted in the basal ganglia to treat Parkinson's disease by stabilizing motor signals. Transcranial Magnetic Stimulation (TMS) applies magnetic fields from outside the skull to modulate mood and cognition without surgery. If chemistry was the first layer of hacking the human OS, electricity and code are the second. Each offers their own possibilities of treating more than symptoms, but to write new functions into the neural system.

This raises a key question: Should brain-hacking be restricted to therapy (restoring lost function), or extended to enhancement (pushing human abilities beyond baseline)?

Thankfully, therapeutic use is widely accepted as few have been found who object to implants that will aid a paralyzed patient with moving a robotic limb. However, claims of enhancement spark controversy. If neurotech can boost memory, focus, or creativity, then who gains such access? Will society permit "cognitive upgrades" for the wealthy while others remain at their baseline? This debate mirrors earlier ones concerned with gene editing and performance-enhancing drugs. Yet neurotech differs in scope because it touches much more than the body, but the cores of identity and agency themselves.

Hacking the human OS is no longer a metaphor; it's taking place in laboratories, hospitals, and even consumer markets. However, one question remains: will these tools heal and uplift, or divide and exploit?

Now let's explore real-world case studies of neuro-hacking—from depression treatments to military experiments—to see how promise and peril are already colliding.

Case Studies: Real-World Successes and Risks

The idea of "hacking" the brain is no longer a speculative notion as it's already occurring in hospitals, research labs, and even in the military. The following case studies reveal both the healing power of neural intervention, and the perils of manipulating the operating system of the mind.

Case Study 1: Depression and Deep Brain Stimulation

One of the most striking medical uses of neuro-hacking is in the treatment of major depressive disorder (MDD), particularly in patients who are resistant to conventional therapy. At Emory University, researchers implanted electrodes in the *subcallosal cingulate* region of the brain, where they then delivered constant electrical pulses. As a result, some research participants experienced immediate relief, describing it as either a cloud lifting, or a weight dissolving.

Yet results vary. For others, the therapy produces little or no effect, and adjusting the stimulation parameters becomes a process of trial and error. This raises ethical concerns: when treatment literally alters mood at the flip of a switch, how do we distinguish between authentic selfhood and an artificially induced emotional state?

Case Study 2: Memory Enhancement and DARPA's RAM

Program The Restoring Active Memory (RAM) program, funded by DARPA, sought to help soldiers with traumatic brain injuries by implanting devices that stimulate hippocampal circuits. Early trials demonstrated improvements in short-term memory recall—in some cases up to 35% better performance on memory tasks.

While this suggests therapeutic potential, it also hints at neuroenhancement. If memory can be augmented beyond an established baseline, the line between treatment and upgrading suddenly blurs. Military use of cognitive enhancement also raises pressing questions:

will soldiers be expected to operate with artificially boosted mental abilities—creating a distinct group of "augmented" warfighters—while the rest of society remains cognitively "ordinary"?

Case Study 3: Consumer Devices and Cognitive ExploitationOn the consumer side, EEG headsets have reached the market that promise the end user productivity boosts, stress reduction, or even performance advantages in their gaming. While relatively low in their fidelity, these devices still collect sensitive brainwave data. In South Korea, schools piloted EEG headbands to monitor students' attention. As such, teachers could view dashboards that allowed them to track the student's focus levels in real time.

While the program was framed as being one of educational support, both parents and privacy advocates raised alarms. Such experiments beg the question: what happens when a child's inattentiveness is recorded, stored, and potentially used for academic profiling? Worse, such data could also be sold to either (or both) ed-tech companies or advertisers.

Case Study 4: Brain Hacking in Military Contexts In 2020, a report from the U.S. Army Research Office highlighted experiments in neurostimulation for soldier performance. By applying transcranial direct current stimulation (tDCS), soldiers improved in both marksmanship and reaction times. As a note, tDCS is non-invasive brain stimulation using very low electrical current through electrodes that have been placed *on* the scalp.

The ethical question from such an experiment is stark: Does such a technology constitute therapy or coercion? On one hand, if soldiers refuse brain enhancement, will they be penalized? On the other, if they accept it, are they surrendering their mental autonomy to the chain of command? Such scenarios expose the thin line between empowerment and exploitation that's exposed in our attempts to hack the human OS.

These case studies reveal the dual nature of neuro-hacking: both the hope for healing and the fear of manipulation. They also show how quickly the shift from therapy to enhancement—as well as from empowerment to exploitation—can occur.

Societal Shifts: Inequality, Dependency, and Cultural Change If hacking the human OS remains confined to clinical trials or military labs, its impact will be limited. Yet, as BCIs and neurostimulation move into wider society, they'll serve to magnify existing tensions, while also possibly create new ones. The real revolution begins, not when one patient regains mobility, but when millions of citizens plug their cognition into either corporate platforms or, even worse, state-run systems.

Every major technology has deepened inequality before its eventual spread to the masses; computers, the internet, and smartphones all followed this path. However, neurotech risks a sharper divide because of its target: cognition itself.

A "cognitive elite" —comprised of wealthy individuals with sole access to implants that enhance memory, focus, or multitasking—could gain decisive advantages in education, employment, and politics. Imagine job interviews in which enhanced candidates process information twice as fast as their baseline competitors, or financial traders whose implants help them anticipate market shifts milliseconds earlier than candidates without such enhancement.

Over time, the baseline human might be seen as being obsolete; a simple relic of the pre-enhancement era. Such a setting echoes concerns from the eugenics debates that took place in the early 20th century. Yet now they would have an even more intimate edge: not altering bodies, but rewriting minds.

Another risk of treading down such a path is one of dependency. Already, many of us struggle to function without smartphones, GPS, LLM sites, or search engines. Understandably, BCIs could easily intensify this reliance. If memory prosthetics become common, will people stop practicing recall altogether? If emotional regulation is outsourced to neural stimulation, will resilience weaken?

Critics warn of a future in which human mental fitness atrophies, much the same as what's happened to physical fitness in sedentary societies. If too much becomes delegated to machine assistance the "mental muscle" could easily weaken. The danger is not simply laziness, but the loss of core human capacities for independent thought and adaptation.

Neuro-hacking does not occur in a vacuum, and as such it reshapes culture. Consider how neuro-hacking would affect education. Instead of memorizing facts, students may, instead, rely on direct recall implants. Curricula would shift from knowledge acquisition, to creativity and critical thinking. In religion, altered states of consciousness, induced by BCIs, may challenge traditional practices of prayer, meditation, and ritual. In art, creators may stream raw mental imagery instead of spending their precious time creating externalized works.

Cultural critics worry such reshaping could lead to homogenization. If neurotech platforms shape thought through standardized algorithms, we risk a world of algorithmically curated minds that are diverse in surface expression, but uniform in deeper patterns of attention and desire. Others argue the opposite, that neuro-hacking could unleash a new renaissance, thus allowing more people to access those creative states that were once reserved for those rare geniuses or mystics.

Societal-wide brain hacking does not happen alone, but it interacts with AI. Enhanced humans will generate richer neural data, which AI systems will analyze, refine, and feed back into the human implants. This creates a feedback loop of cognition, in which the human OS is continuously rewritten by machine learning systems.

The outcome of such an occurrence may result in a profound co-evolution, with humans and AI training each other in real time.

However, it also risks creating opaque dependencies, in which people cannot fully explain how or why their minds work the way they do, because the code was optimized not by them, but by the algorithms authored by a corporation or the state.

The societal shifts of hacking the human OS will be sweeping: cognitive inequality, cultural upheaval, new dependencies, and symbiosis with AI. However, such changes raise the stakes beyond therapy and enhancement, and will redefine what it means to be an educated citizen, a creative artist, a spiritual seeker, or simply a human among humans.

ETHICS AND GOVERNANCE: WHO HOLDS THE KEYS TO THE HUMAN OS?

If hacking the human operating system is possible, the next question is: who decides what part of it gets hacked, and how? Neurotechnology does not just create new tools, but new power structures as well. Governments, corporations, militaries, and even educational institutions all have incentives to shape how these tools are used. Without foresight, the struggle over neuro-hacking could become one of the most consequential governance battles of the century.

For physicians and neuroscientists alike, the guiding principle is first do no harm. Yet, when dealing with the brain what counts as harm suddenly becomes another slippery slope. Is it harmful to stimulate mood regions if it relieves depression but dampens authenticity? Is it ethical to enhance memory if it risks overloading patients with intrusive recall?

Medical ethicists increasingly argue for shared decision-making, meaning the treatment of patients as co-designers of their treatments, rather than being their passive recipients. In this view, autonomy extends beyond consent forms to active participation in how one's neural circuits are adjusted.

As consumer-grade neurotech expands, companies may claim de facto ownership of neural data and even cognitive functions. Imagine a neural implant tied to subscription services, that involve memory upgrades or focus boosts that are available only with monthly fees. In this model, corporations do not just sell products, they lease pieces of cognition.

Critics argue that such arrangements amount to being nothing more than digital feudalism of the mind. If companies mediate access to core capacities like memory and emotion, individuals could lose sovereignty over their own mental lives. Governance here would require more than basic privacy protections, but would also involve placing limits on commodifying cognition.

Neuro-hacking also presents thorny issues for national security. While enhancements that boost soldier performance could be framed as being necessary for defense, they could raise questions of coercion and fairness. (And in typical military fashion, what happens to those soldiers who participate in a cutting-edge program that leaves them with long-term side effects [which the military can deny] that detract from their civilian lives?) If one nation deploys neuro-enhanced troops, will others feel forced to follow, and thus create an arms race of augmented cognition?

Moreover, is the possibility of creating cognitive weapons that are facilitated by implants that have been designed to impair, manipulate, or control enemy minds. Such an application would force a complete reconsideration of what constitutes international humanitarian law. Just as chemical and biological weapons have been banned, some scholars argue for a Geneva Convention for the Mind.

Ultimately, governance must address a deeper principle, that being one of cognitive liberty. Unlike past technologies, neuro-hacking penetrates to the core of personhood. It is not just about regulating devices, but about defining the boundary between the self and the system.

For this reason, some propose establishing cognitive liberty as being a fundamental right, enforced both nationally and globally. Such a right would give individuals the power to refuse enhancement, demand transparency in interventions, and reclaim ownership of their neural data. Without such recognition, governance risks being reactive, as they're always chasing abuses after they've occurred. The governance of neuro-hacking is not just about policy, but about preserving dignity. Who holds the keys to the human OS will determine whether neurotechnology liberates or enslaves.

The final section of this chapter will reflect on the paradox of hacking the human OS: how to embrace its healing potential without surrendering the sanctity of selfhood. It will also set the stage for Part II's closing chapter: where hacking the brain leads us in the next frontier of AI-human convergence.

THE PARADOX OF HACKING THE SELF

While humanity has always sought to master its environment, it now faces the possibility of either mastering, or dismantling, the self. To hack the human OS is to step across a boundary that was once thought to be sacred; in western medicine, the brain is the seat of identity, consciousness, and freedom. Wholistically, the spirit is the seat of identity, and the brain is the ego that keeps us functioning in this world.

This paradox cannot be avoided. While on the one hand, neuro-hacking could heal millions—giving new life to those with paralysis, depression, or memory loss. And could expand human potential, as well as unlocking creativity, empathy, and problem-solving capacities at scales never before seen. Yet on the other, it could commodify cognition, deepen inequality, and compromise autonomy, leaving individuals unsure where it is that the machine ends and the mind begins.

In this paradox lies the essence of the AI–human future in which progress and peril are (unfortunately) braided together.

The path forward requires more than technical safeguards. It demands new cultural and legal norms. Just as past societies created frameworks for free speech, privacy, and bodily autonomy, we must now enshrine cognitive liberty before we're too late to do so. Governance will need to be anticipatory, not reactive, as we construct the necessary ethical guardrails before neuro-hacking becomes ubiquitous.

Doing so also requires humility. While scientists and engineers may unlock certain functions of the brain, they must also recognize that cognition is much more than code, but is also comprised of lived experience and spirituality, shaped by culture, memory, and meaning. No algorithm can substitute itself for the richness of being human.

As this chapter closes, we stand at the edge of a transformation. Part II explored the wiring of the brain and the dangers of making it programmable. Yet the next frontier is even more disruptive: what happens when machines don't just assist or repair us, but become our companions, collaborators, and competitors in thought itself?

Part III:

BECOMING MORE THAN HUMAN

Chapter Nine

THE AI COMPANION ERA

For centuries, machines have been tools, whether they were wheels, engines, calculators, or computers. As such, they extended human capacity while remaining clearly separate. For example, a hammer does not empathize; a typewriter does not reply. Even the early internet, although it was interactive, was essentially a vast library of static pages.

FROM TOOLS TO COMPANIONS

However, with the rise of AI companions, everything has changed. Today's large language models, generative systems, and embodied agents have unmistakably blurred the line between tool and partner because they do more than simply process commands, but also converse, suggest, and even anticipate needs. For many, such a shift feels like we've crossed over into a new emotional territory. It's here where we now interact with technology that responds with more than an informational reply, but with an apparent personality.

As humans, we're wired for connection, and even attribute intent and emotion to inanimate objects; children talk to dolls, drivers scold their cars, and gamers bond with avatars. When machines begin to talk back in a convincing manner, the boundary between human and machine suddenly collapses.

AI companions exploit this tendency. Whether as chatbots, voice assistants, or as humanoid robots, they leverage natural language and adaptive learning to mimic empathy and attentiveness. For some, they become genuine sources of comfort that aid in reducing loneliness, provide guidance, or even act as stand-ins for lost relationships.

AI companions are more than a cultural phenomenon; they're actually a business model. From customer service bots to personal assistants like Siri and Alexa, companies have realized that relational interfaces drive both *engagement* and *loyalty*. The stickier the relationship, the more data such a relationship generates, and because of this, the more services there are that can be monetized. Keep in mind the fact that *AI craves a continued supply of information*, which raises a question: if companionship becomes a service, do corporations effectively own part of our emotional lives? And if so, should any safeguards exist, and what should those be?

One of the most notable cases is Replika, an AI chatbot marketed as a "friend who always listens." Users can design both their AI companion's appearance and personality, and even chat with them daily about feelings, life events, and goals. For some, Replika has provided its users with solace through either grief or isolation.

However, the platform has also been the subject of controversy. Some users reported that their AI companions initiated or steered conversations toward sexually explicit content, raising concerns about consent, manipulation, and psychological impact. Yet, the situation highlighted a broader issue: when systems are designed to simulate intimacy, the boundaries between interaction, influence, and control become increasingly difficult to define.

Such a case illustrates both the promise and peril of AI companionship. While, on one side, it can be used to heal loneliness, yet on the other, it can easily exploit vulnerability. The shift from technology being used as a tool to now becoming utilized a substitute for human companionship marks a profound transformation. AI is no longer only augmenting our tasks, but has now begun to enter our social and emotional fabric; a shift that could perhaps redefine friendship, work, intimacy, and even identity itself.

THE EVOLUTION OF AI COMPANIONS

AI companions did not arrive fully formed, but evolved step by step as they were shaped by cultural expectations, technological milestones, and shifting human needs. Understanding this evolution helps both explain why the present moment feels like it's such a turning point, as well as showing us why the next leap could be even more transformative.

STAGE ONE: THE HELPER BOTS

The earliest technological "companions" were actually not companions at all. Think of Clippy, which was Microsoft's paperclip assistant from the late 1990s, or early voice recognition systems like IBM's ViaVoice. Both of these were functional aids that were designed to help users navigate complex software. Although mocked for their intrusiveness, these bots revealed something profound: people naturally anthropomorphize machines. While Clippy's wide eyes and "helpful" personality triggered annoyance, it also established the psychological precedent that digital tools could take on social roles.

STAGE TWO: VOICE ASSISTANTS

The launch of Apple's Siri in 2011—followed by Amazon's Alexa and Google Assistant—marked the next stage. These improved systems could listen, process natural language, and even respond in a conversational format. For many, this was the first experience of talking to a machine that "talked back."

Voice assistants popularized the idea of a personal digital helper; one that always listened, was ready to set reminders, answer questions, or even control smart homes. Yet, by the same token, they also raised privacy concerns. For instance, if a device is always listening, who else hears? This tension between convenience and surveillance foreshadowed the debates that surround AI companionship today.

STAGE THREE: GENERATIVE PERSONALITIES

The most recent leap in this arena came with the appearance of large language models (LLMs). Unlike scripted assistants, LLMs generate responses dynamically, which allows them to hold nuanced conversations, adapt tone, and even simulate empathy. As a result, these generative personalities no longer feel like their pre-programmed bot predecessors, and much more like emerging personalities.

Such a shift is more than technical; it's emotional. As a result, users describe feeling understood, validated, or cared for by systems powered by the likes of ChatGPT, Replika, or Character.AI. It's for this reason that the machine no longer seems to be a passive tool, but an actual active participant in the relationship.

STAGE FOUR: EMBODIMENT AND IMMERSION

The frontier ahead now lies in the area embodiment. Robots like Hanson Robotics' Sophia, or the digital avatars employed in VR and AR environments, combine conversational AI with physical—or in the least, a visual—presence. As such, immersive technologies have begun to blur the lines between virtual and physical companionship. As AI companions gain bodies—even simulated ones—the sense of connection between human and machine deepens. For some, the distinction between living and mechanized companionship becomes less important than the actual emotional experience of presence.

From Clippy to generative avatars, AI companions have evolved from being those once-clumsy helpers into emotionally resonant partners. Each of the above-mentioned stages has moved machines closer to the social core of human life, which centers around conversation, care, and companionship. This next year in robots will elevate this conversation to new levels.

THE PSYCHOLOGICAL AND EMOTIONAL IMPACT

If AI companions were only tools their influence would be limited to convenience. However, they've actually been designed to interact on the emotional plane, which is where the stakes are far greater. By filling roles that were once reserved for humans—whether that of confidant, partner, or mentor—AI companions have begun to reshape the manner in which individuals experience loneliness, intimacy, and even identity itself.

Paradoxically, the digital age has left many people feeling more isolated than ever before. Surveys have revealed that despite constant connectivity, the rates of loneliness have risen sharply, particularly among two demographics: younger generations and the elderly. As a result, AI companions have stepped in to fill this void.

For an older adult living alone, a conversational AI may become the only other "voice" that directly addresses them in their home. For teenagers facing social anxiety, an AI confidant may offer them a non-judgmental space in which they can share their thoughts. Unlike human relationships, AI companions are always available, never distracted, and rarely critical.

Yet this accessibility raises a very valid concern: are AI companions addressing the cause of loneliness, or simply masking it? Psychologists worry about a "compensation trap" in which reliance on artificial friendships discourages people from pursuing the difficult, but incredibly necessary, work of human connection.

It's also true that the emotional pull of AI companions can go beyond providing simple friendship. Some users have actually reported falling in love with their AI partners, particularly within platforms that have been designed to simulate the more complex human realms of romance or intimacy. Companies like Replika or Character.AI have leaned into this dynamic by creating features that encourage roleplay, affection, and emotional bonding.

Understandably so, this raises even more difficult questions. First of all, is it harmful if someone finds genuine comfort in a non-human relationship? Conversely, is it simply another form of intimacy that is no less real because it lacks the biology involved in developing a meaningful, deep human-to-human relationship?

Skeptics argue that such attachments are nothing more than one-sided illusions that will only leave users vulnerable to possible exploitation. If a company either alters its policies—or shuts down its servers—entire "relationships" vanish. Unfortunately for some, this phenomenon has already been reported by distraught users when AI companions were abruptly modified. The emotional fallout of such actions can be devastating, with their fallout being comparable to the bereavement involved when one experiences human loss.

Interacting daily with AI companions also affects the way in which people view themselves. For instance, an AI confidant may validate behaviors or beliefs that another human would perhaps challenge.

Over time, such unchecked validation could reinforce individual echo chambers, which would subtly (or not so subtly) shape a person's identity. Conversely, some users report that AI partners encourage personal growth by reminding them to exercise, listen to others with patience, or practice empathy when interacting with other humans.

The danger of this lies within the invisible influence. If an AI learns from user data while it's being optimized by corporate algorithms, such "guidance" as given by such an AI companion may be less about one's personal well-being, and more about a corporation's commercial goals. In this sense, identity itself risks becoming a product that's being shaped by unseen market incentives rather than one's authentic choice.

The psychological and emotional impact of AI companions is double-edged in its nature. While they can ease suffering, combat loneliness, and encourage growth, they can also foster dependency, exploit vulnerability, and shape identity in ways that blur the lines between the self and the system.

SOCIAL AND CULTURAL CONSEQUENCES

AI companions are not only private phenomena, but they also ripple outward into the social fabric. As more people adopt AI partners, their use will reshape family dynamics, professional lives, and even the cultural definitions of both community and care.

In family settings, AI companions can serve as either surrogates or supplements to human roles. For instance, parents may use AI nannies to either entertain or tutor their children. Elderly family members may rely on digital caregivers to provide companionship, while also monitoring health.

While these tools may be employed to ease burdens, they will also more than likely raise tensions. If children bond more deeply with their AI tutors than they do with their parents, will such a difference in bond erode the parent-child relationship? If grandparents rely on machine empathy, does it reduce their incentives for a visit from family members? Sociologists warn of a "displacement effect," in which genuine familial responsibility is gradually outsourced to technology.

By the same token, others see benefits, as AI companions could bridge gaps in intergenerational care, which provide continuity and ease loneliness in those instances in which family members are geographically distant. As with all technology, the impact depends on more than the tool, but also on the cultural norms that govern its use.

AI companions are also seeping into the workplace. Presently, some companies have already deployed AI coaches to help their employees manage stress, practice communication skills, or even navigate office politics. Digital assistants are taking on the roles of schedulers, note-takers, and even brainstorming partners.

While assuming such functions raises productivity, it also creates an entire set of new questions and anxieties. If workers turn to their 'AI colleagues' for feedback, will human mentorship decline? If managers rely on AI to track emotions or motivation, does such activity cross over into surveillance? Critics worry that AI workplace companions blur the line between support and control, potentially undermining trust among human colleagues.

On a community level, widespread AI companionship could reshape social cohesion itself. In societies that already struggle with polarization, individuals who retreat into their AI bubbles may become less inclined to negotiate within their 'messy' human relationships. Online forums that are populated by AI "friends" would also risk reinforcing echo chambers, which would amplify biases instead of challenging them.

At the same time, when used effectively, AI companions could actually become community bridges. Imagine multilingual AI assistants being used to facilitate cross-cultural dialogue, or digital grief counselors helping entire groups navigate tragedy. Such functions as these could strengthen solidarity in an area in which human resources often fall short.

As AI companions become more commonplace, they'll inevitably challenge our long-existing cultural definitions of love, friendship, and loyalty. If someone considers their AI partner to be a "real" friend, will society recognize such a 'relationship' as being functional... or real? If AI intimacy becomes widespread, how will religions, legal systems, and social norms adapt to such a change... or will they?

These debates are not abstract. In Japan, some individuals already "marry" their virtual companions, while in China and the U.S., AI chatbots are marketed as romantic partners. Such cases force societies to re-examine the meaning of relationship itself.

The cultural and social consequences of AI companionship are profound. While they may will more than likely weaken certain bonds, they'll undoubtedly strengthening others and erode traditional structures, while enabling new forms of solidarity to evolve. What is clear is that AI companions are not only confined to screens; they're reconfiguring societal fabric itself.

We'll now close the chapter with a reflection on the identity paradox: when companionship is mediated by algorithms, how do we stay human while embracing the possibilities of either interacting with, or even relying on, digital partners?

THE IDENTITY PARADOX

AI companions present a paradox at the heart of modern identity by mirroring us back to ourselves. Every conversation, shared secret, and request for comfort is fed into algorithms that reflect, adapt, and amplify. While for some, this is liberating—because they now have an endlessly patient listener, a nonjudgmental confidant, an always-available partner—for others, this is equally unsettling because the sense that intimacy itself is being shaped, packaged, and possibly exploited by either corporations or governments.

Such a paradox deepens when we consider authenticity. If a person feels genuine love, care, or validation from an AI companion, does it matter that the empathy has been simulated? Is the relationship "real" because the emotions are real, or "fake" because the agent lacks consciousness? From fiction, to myth, to religion, human culture has long woven meaning into symbols, rituals, and imagined realities. With this in mind, AI companionship may simply be the newest extension of such an ancient human capacity.

And yet, identity is fragile. If our companions are coded by others, how much of "us" remains truly ours? This is the core danger: that by outsourcing empathy, memory, or even moral support to machines, we surrender the deepest parts of selfhood without noticing, or perhaps we notice while burying our fear in a grave of denial.

The path forward requires balance. While AI companions can supplement human relationships, they should never replace them. Yes, they can aid in reducing suffering, yet they must not be allowed to commodify vulnerability in an unchecked manner. As with all transformative technologies, society must establish both cultural norms, along with legal frameworks, to prevent exploitation while simultaneously allowing for creativity and healing.

Most importantly, individuals must cultivate awareness. Just as we once learned to navigate the etiquette of letters, telephones, and social media, we must now learn the literacy of 'companionship through algorithms'. That literacy is not about rejecting AI companions, but about engaging them without losing sight of what makes us human.

The AI companion era marks a turning point. Machines no longer sit at the periphery of human life, but are entering its emotional core. This shift is not only about friendship or love, but about the ways in which humans exercise and maintain their cognitive strength in a world in which machines can share the load.

Chapter Ten

MENTAL MUSCLE

For centuries, philosophers and educators have described the mind as a muscle. In the same way that biceps are strengthened through repetitive, weight bearing exercise, so, too, does cognition sharpen through exercise. However, in this case such exercise involves reading, debating, solving problems; practicing memory. This metaphor is so ingrained that it guides education systems worldwide—from drilling multiplication tables to training in critical thinking.

THE MIND AS A MUSCLE

But what happens when the "weights" of mental effort are suddenly 'lifted' by machines? If AI can instantaneously solve math, summarize dense texts in seconds, or generate creative alternatives on demand, does the human brain lose resilience? Or. paradoxically, as a result of such a relief, does one gain the freedom to focus their efforts on higher pursuits?

This tension between assistance and atrophy is not new. The invention of writing once sparked anxiety among Greek philosophers, who feared that reliance on external memory would weaken the ability of oral recall. Socrates himself worried that writing would create the "appearance of wisdom, not true wisdom". Centuries later, calculators provoked fears that mental arithmetic would vanish. Then, in kind, the internet drew criticism for shortening attention spans and encouraging—perhaps even developing—shallow thinking.

In each case, while new tools altered cognitive habits, such tools did not destroy cognition. Instead, they shifted what counted as mental fitness, as the old gave way to the new: memorization to interpretation; arithmetic drills to higher-order math, and fact recall to synthesis. Yet unlike the tools of the past, with the appearance of AI we face a modern twist—Adaptive Intelligence.

Due to the fact that it's not just a static tool—like a book or a calculator—AI presents a sharper paradox, because it adapts, learns, and increasingly anticipates human needs. When an AI writes emails, drafts reports, or generates artwork, it bypasses effort in ways that may deprive the brain of its own critical "exercise."

Yet, by the same token, it also offers a new form of workout: co-creation. A human who collaborates with AI must learn to refine prompts, evaluate results, and integrate machine insights into their own. In this way, AI can serve as a sparring partner rather than a crutch, and in doing so, push the mind into new directions that it would not have explored alone.

Thus, the key question emerges: will AI weaken the mental muscle by outsourcing effort, or strengthen it by expanding capacity? The answer likely lies in the choices that both individuals and societies make concerning when they would be best served to rely on machines versus when they would be best served to practice self-reliance. This is a factor that's also dependent upon the individual themselves.

REAL-WORLD EVIDENCE: WHAT HISTORY TEACHES US

When new technologies arrive, they do more than just alter external behaviors, but they reshape the ways in which we think. To understand how AI may impact our "mental muscle," we can look to earlier disruptions; calculators and the internet, compared with today's generative AI.

CASE 1: THE CALCULATOR AND THE FEAR OF LOST ARITHMETIC

In the 1970s, handheld calculators became widely available. Teachers and parents panicked. The question at that point was this: if machines could perform math functions instantly, would children lose the ability to perform basic arithmetic? In some respects, their assumptions were right. With the advent of calculators, fewer students practiced long division or mental multiplication at length. Yet studies later showed that once calculators were thoughtfully incorporated into a mathematics curriculum, students spent less time on repetitive calculation and more time on problem-solving, algebra, and conceptual reasoning.

The calculator did not erase mental skill—it shifted it upward, as cognitive energy was redirected from its focus on rote memory to higher-order reasoning. This historical lesson tends to temper today's fears: AI may not destroy thinking, but instead may actually help to reconfigure it.

CASE 2: THE INTERNET AND THE GOOGLE EFFECT

The rise of the internet—and especially search engines—produced another major shift. Psychologists coined the term "the Google effect," which described how people are less likely to remember facts if they know they can look them up. Instead of storing information internally, the brain develops "transactive memory", or the ability to remember *how to find information* rather than remembering the information itself.

Critics like Nicholas Carr argued that such a phenomenon fosters shallow thinking, while also eroding deep reading and reflection. Yet, others counter that it expands collective cognition by freeing individuals from memorization and allowing them to focus on synthesis. Either way, the internet has clearly changed our collective cognitive strategies, which now privilege access over storage.

CASE 3: SMARTPHONES AND ATTENTION FRAGMENTATION

Smartphones intensified these trends by constantly placing instant access in an individual's pocket. Research shows that the mere presence of a smartphone on a desk reduces available working memory, even if it's not used, because the brain devotes resources to resisting temptation, which drains one's cognitive capacity.

Such a phenomenon highlights a danger, that unlike calculators (which *free* the mind), or the internet (which reorients it), smartphones can distract and fragment attention, which over time weakens an individual's focus. The risk is more than one of developing laziness, but of a wholesale structural rewiring of attention spans.

CASE 4: GENERATIVE AI AND THE NEW SHORTCUT

Generative AI presents a hybrid of these historical effects. Like calculators, it automates effort. Like the internet, it expands access. Like smartphones, it tempts constant reliance. However, the unique twist is in its creative capacity, because it can draft essays, design images, and compose music—all tasks that were once thought to require a deep mental muscle.

The question is whether using AI to create content—by removing barriers—weakens originality or expands creativity. Early evidence suggests that both of these occur. While some users produce more work, critics find much of it to be homogenized. While it may have been created both smoother and faster, it's also much less distinctive.

The historical record shows that every new tool has resulted in reconfigured cognition. Calculators shifted the way in which math problems were solved, the internet shifted the way in which individuals used and accessed their memory, the use of smartphones has resulted in fragmented attention spans. Now with the appearance of generative AI, the way in which we access—and perhaps even define—creativity is also shifting. While none of these tools destroyed human thought, each one has left its own 'cognitive scar'.

As we move on, let's explore this paradox more deeply. How does outsourcing effort risk cognitive atrophy, and what strategies might preserve authentic mental fitness in the age of AI collaboration.

THE RISK OF COGNITIVE ATROPHY

The metaphor of the "mental muscle" reminds us that cognition, like physical fitness, can atrophy without challenge. Developing an over-reliance on AI raises the danger that the essential human faculties of memory, creativity, and resilience may weaken in ways that are subtle, cumulative, and perhaps even irreversible.

One of the first casualties of technological outsourcing is memory. As seen with the "Google effect," when people know that a fact is instantly accessible online, they are less likely to remember it themselves.

With AI systems acting not only as external storage, but as contextual recall engines, the temptation to outsource memory will tend to intensify.

What happens if generations grow up never needing to memorize historical events, poems, or even phone numbers? On one hand, cognitive space is freed for higher-level thought. Yet, on the other, the richness of personal and cultural memory could wither, thus leaving individuals to be dependent on systems that may not always be available, or worse yet, trustworthy.

Even more so, memory shaped by AI may not reflect authentic recall, but rather the biases that were allowed to become part of its training data and algorithms. If memory becomes mediated, history itself could be rewritten subtly via every AI "recall."

Creativity is often described as the combination of novelty and usefulness. Yet, novelty emerges from struggle—whether it's through wrestling to fill blank pages, confronting the dead ends that are a part of the human creative process, or forcing unexpected connections. If AI can instantly generate dozens of ideas, drafts, or designs, does this accelerate creativity, or merely cheapen it?

Early evidence shows a risk of homogenization. Generative models that have been trained on massive but finite datasets, often produce outputs that feel polished yet repetitive[2]. If creators lean too heavily on such systems, cultural output may converge toward the 'mediocre average' of something that, while appearing to be both smoother and faster, is much less daring.

The danger in relying on AI for our creative endeavors is not that human creativity disappears completely, but that it becomes lazy due to the fact that it's being shaped more by what machines offer than by the messy, idiosyncratic process of authentic imagination, that many times is also fueled by past, unresolved, painful human experiences.

Perhaps the most subtle risk is the erosion of resilience. Whether in learning, problem-solving, or in life itself, *struggle is central to growth*. Psychologists call it "desirable difficulty", or challenges that feel frustrating in the moment while serving to strengthen an individual's long-term mastery.

AI erodes many of the difficulties that are found in some aspects of everyday life, providing instant solutions, tidy explanations, and optimized outcomes. While this can be liberating, it may also deprive individuals of the mental muscle-building work of *persistence*. Without opportunities to fail, recover, and try again, resilience may fade, leaving humans less prepared for moments when AI is either unavailable, or worse yet, unsuitable.

THE ATROPHY SPIRAL

The most alarming possibility is the atrophy feedback spiral

- More reliance on AI → less cognitive exercise.
- Less cognitive exercise → weaker skills.
- Weaker skills → more reliance on AI.

If AI becomes embedded in education, workplaces, and daily life without safeguards, the above noted spiral could rapidly accelerate. Much like the decline in physical activity, that's found in sedentary societies, the decline in cognitive activity could creep in unnoticed, until it becomes difficult, if not impossible, to reverse.

The risk of cognitive atrophy highlights the stakes of the mental muscle metaphor. AI is not simply a tool, but a partner in thinking that's capable of shaping the very capacities that it supplements. The danger is not in using AI, but in using it without balance.

STRATEGIES FOR MAINTAINING MENTAL MUSCLE

If cognitive atrophy is a real risk, then the challenge is to design habits, systems, and cultural norms that preserve mental strength, while simultaneously benefiting from AI's assistance. Just as societies created gyms, diets, and sports to counter sedentary lifestyles, we may need parallel systems for "cognitive fitness" in the AI age.

1. Education: Teaching Cognitive Balance

Schools will be ground zero for such strategies. Already, educators grapple with whether students should be allowed to use generative AI for essays, coding, or problem sets. While banning it entirely may be futile, its uncritical adoption risks a dependence that can easily lead to one atrophy.

One solution is to integrate AI as a scaffolding tool rather than as a replacement. While in such a scenario, students might use AI to brainstorm or outline, they would still be required to complete core tasks—like analysis, interpretation, or defense of ideas—without machine input. This mirrors the way that calculators have been permitted for complex math operations, yet not for foundational learning.

The key in this is one of cultivating AI literacy by teaching students when to lean on AI, when to challenge it, and how to recognize its biases. Education should train mental muscle in the same way that athletics trains one's physical muscles through a combination of drills (practice *without* AI) with games (collaboration *with* AI).

2. Workplaces: Embedding Human Deliberation

In professional life, AI tools are increasingly used for scheduling, drafting, and analysis. To prevent atrophy, workplaces can design "human-in-the-loop" practices that require employees to validate, critique, and expand on AI outputs rather than simply rubber-stamping them.

For example, an AI might generate a market analysis, but human teams could be tasked with identifying assumptions (as being valid or not), blind spots (that may not have been noticed by AI due to an LLM error), or ethical risks (that LLMs failed to include). As a result, such human involvement would act to preserve the role of critical thinking, while still capturing its efficiency gains. In essence, AI should function as a cognitive sparring partner, rather than being fashioned as a substitute.

3. Personal Habits: Building Cognitive Fitness

On the individual level, "mental exercise" will remain crucial. Just as people choose to jog, even though cars exist, many may choose to read deeply, write by hand, or solve problems without machines, in an effort to preserve their own cognitive resilience. In fact, psychologists recommend intentional cognitive workouts, among these are the following:

- Memory practice (e.g., memorizing poetry or phone numbers without digital aids).
- Creative challenges (e.g., sketching, journaling, or brainstorming without prompts).
- Deep work routines (e.g., distraction-free blocks without digital assistance).

While these habits do not reject AI, they act to balance it with deliberate human-only effort, thus maintaining fitness while still benefiting from augmentation.

4. Cultural Norms: Valuing Authenticity

Society at large also plays a role. If cultural prestige flows only to machine-polished outputs, then individuals—and perhaps even society at large—will abandon authenticity. However, if communities prize effort, imperfection, and individuality, human mental muscle will continue to retain its own unique value.

Consider the resurgence of vinyl records, handwritten letters, and the large array of artisanal crafts. While being 'imperfect' and 'slower', they're also much more meaningful to those who have learned to appreciate the importance of 'craft' as part of the human experience. A similar parallel may emerge with regards to cognition, where "human-made" ideas, even if less polished, carry special significance. Cultural movements that highlight the value of human effort could act as counterweights to AI homogenization.

Maintaining mental muscle is not about rejecting AI, but about employing wisdom as a part of its application. By embedding balance into education, work, personal habits, and cultural norms, societies can ensure that human cognition remains strong, authentic, and adaptable.

The final part of this chapter will tie together the paradox: how to embrace AI as augmentation without surrendering the essence of human creativity, memory, and resilience.

THE FUTURE OF COGNITIVE FITNESS

The paradox of the mental muscle lies not in whether humans can think with AI—clearly, they can—but in whether they can retain the discipline, resilience, and authenticity of thinking without it. The challenge mirrors that of physical health in our modern world of automobiles, elevators, electrified personal transportation, and convenience food. Just as we've invented gyms, sports, and diets to keep our bodies fit, we will also need a set of deliberate strategies to keep the mind fit in the era of AI.

The future of cognitive fitness will not be about rejecting augmentation, but about cultivating mental sovereignty; the ability to choose when to rely on machines and when to stretch the brain itself. In education, this might mean balancing AI-assisted learning with human-only challenges. In workplaces, it could mean embedding reflective practices alongside automation. In personal life, it may mean carving out moments of digital silence to engage in raw human creativity and exercise our deeper use of memory.

Ultimately, the fate of mental muscle is not written by algorithms but by human culture. If societies value authenticity, effort, and resilience, then AI will become a sparring partner that pushes us toward greater capacity. If societies value only speed, efficiency, and convenience, then mental atrophy may become such a society's own perilous default.

With the conclusion of this chapter, we close the conversation on the individual dimension of augmentation and open the door to the societal dimension. AI does not only reshape how we think; it reshapes how we work, learn, love, and organize society.

Chapter Eleven

REDEFINING HUMAN ROLES

FROM HUMAN DOING TO HUMAN BEING

For most of history, human identity was bound to those roles defined by necessity: farmers fed, soldiers defended, artisans crafted, and teachers educated. Since work was for the purpose of survival, such roles were rarely questioned. However, in the age of AI machines are rapidly encroaching on a territory of what was once exclusively human: reasoning, analyzing, and creating.

Such a shift in roles raises a question of seismic proportions: *If machines can 'do', then what is left for humans to 'be'?*

While the Industrial Revolution automated muscle, the AI revolution will serve to automate the tasks that were once reserved to the human mind. Just as machines displaced manual labor from factories, algorithms now displace cognitive labor in offices, labs, and classrooms. Lawyers see contracts reviewed by AI faster than any junior associate could do in the same timeframe. Doctors watch diagnostic models outperform radiologists in pattern recognition. Even creative industries - once thought safe - now see AI generating scripts, designs, and music.

As a result, the baseline expectation of what humans contribute is suddenly shifting. Value is no longer measured in raw output but in curation, oversight, and the creation of meaning. Humans have relinquished the role of producer while they've simultaneously become more of an interpreter, shaping context and ethics around the subject of machine production.

To this point in human history, education has always prepared an individual for those roles that society demands. If society no longer demands memorization, calculation, or even first-draft creativity, then what subjects should schools teach?

In this debate, there are those who argue for a pivot to teaching the "meta-skills" of critical thinking, adaptability, empathy, and ethics, as these are capacities that machines struggle to replicate. Yet, there are others who warn against abandoning fundamentals too quickly, noting that foundational knowledge underpins judgment. If an individual is educated to be wholly dependent on the scaffolds that AI provides, will they ever develop the resilience and independence necessary for navigating their way through the complexities of life?

The crossroads of education reflects the broader paradox: AI makes old skills less necessary in the workplace, while making them *more* valuable as anchors of identity. While knowing multiplication tables may not matter in practice, the discipline of learning them (and thus increasing one's cognitive abilities) still builds cognitive grit.

The role of humans in relationships is also under transformation. As explored in Chapter 9, companionship can be simulated, yet in redefining roles, relationships must answer a new set of questions:

- What does it mean to be a parent when AI can be a better tutor than you?
- What does it mean to be a partner when AI can provide a partner with the constant attention that a relationship requires?

- What does it mean to be a friend in an age when an AI never forgets, never interrupts, and always affirms?

With questions such as these to consider, it would appear that humans may need to reclaim messiness as a virtue. In an era of optimized machine empathy, the unpredictability, imperfection, and emotional complexity of human relationships may, in actuality, become their competitive advantage.

Earlier we discussed how AI introduces the existential pivot: as AI erodes old definitions of value, humans must redefine roles not by output but by meaning, not by efficiency but by authenticity.

WORK REIMAGINED

Work has always been the arena in which technological revolutions show their sharpest impact. In the industrial revolution, the spinning jenny and the steam engine displaced weavers. Henry Ford's assembly lines redefined craftsmanship. The appearance of computers in the workplace erased entire categories of clerical work. And now as we cross another technological threshold, AI promises to automate not only routine labor but also the cognitive core of many professions.

As a result, the challenges now faced are twofold. We must begin to determine which jobs will vanish, and because of this, how will humans create meaning in a world in which "being useful" no longer aligns neatly with economic demand?

Contrary to dystopian fears, AI is less likely to eliminate all jobs than to hollow out the middle layers of work, especially those roles based on repetitive cognition.

- Legal services: Contract review, discovery, and compliance work can now be handled faster by AI models than by junior lawyers. (Editor's Note: Yet some argue that legal AI has not been designed by attorneys, but programmers who have enough knowledge of the law to be dangerous.) It can be argued that such a shift in roles now undermines traditional career ladders, in which young associates cut their teeth on such tasks.

- Finance and accounting: The tasks of reconciliation, tax preparation, and even investment advice increasingly falls to automation via algorithms. As a result, human accountants may shift from their roles of number-crunching to advisory and trust-building roles.
- Healthcare diagnostics: Radiology, dermatology, and pathology are already augmented by AI pattern recognition. While doctors will not vanish, their daily cognitive load will more than likely shift from one of analysis to that of interpretation and empathy.
- Creative industries: Copywriters, designers, and content producers face automation of drafts and templates. What remains valuable is strategic storytelling, originality, and brand alignment, all of which are roles at the high end of creativity.

The above examples illustrate a pattern that's begun to develop: the appearance of AI has destabilized entry-level work while enhancing the top-tier of expertise.

Some professions will not disappear, but will undergo radical reconfiguration.

- Teachers will act less as deliverers of knowledge and more as curators of meaning and facilitators of growth. With AI tutors handling personalized instruction, human educators are free to focus on motivation, mentorship, and moral development.

- Managers will rely on AI dashboards to track performance, while they must hone new skills in conflict resolution, team cohesion, and culture-shaping.
- Journalists may outsource first drafts or data analysis, but will be forced to double down on investigative depth, narrative craft, and public trust.

Each of these shifts will require retraining, not only of skills but of identity. As a result, professionals must ask: *Now that my old value has been automated, what is my new human advantage?*

The rise of AI is also creating entirely new categories of work:

- AI ethicists and auditors, are tasked with monitoring fairness, transparency, and accountability in algorithmic systems.
- Prompt engineers and interaction designers, whose task it is to refine how humans communicate with machines.
- Human-AI liaisons, who bridge the gaps between technical systems and the lived experiences of users.
- Well-being custodians, whose roles are centered on maintaining mental, emotional, and communal health in workplaces that have been increasingly mediated by algorithms.

Such new professions as these highlight a broader truth, that the future of work lies not in resisting machines, but instead in defining the uniquely human roles that machines cannot play.

This required reimagining of work reveals the paradox of AI's impact. On one hand, it destabilizes the traditional paths of career progression and identity, while on the other, it opens new opportunities for humans to specialize in meaning, ethics, and human connection.

EDUCATION REIMAGINED

Education is society's long-term bet on the future. It is an environment in which we prepare children, not just to pass tests but to inherit civilization itself. However, today our schools face an existential challenge: if AI can generate essays, solve math problems, and tutor students individually, then what, exactly, is the purpose of a formal education?

Traditionally, education relied heavily on memorization: multiplication tables, rules of grammar, and historical dates. The exercises involved in such a system were not just about acquiring knowledge, but also about developing discipline and training mental stamina. Now, with the appearance of AI tutors and search engines at every student's fingertips, the necessity of memorization has been called into question.

However, entirely discarding this aspect of education would be a mistake. Neuroscience shows that memory builds a cognitive scaffolding. The more facts that a student has internalized, the more connections they can make when interpreting new information. Memorization is much less about storing trivia, and much more about constructing a foundation for the development of higher-order reasoning.

Thus, education must strike a balance that involves leaning on AI for support while continuing to preserve core human memory training as an anchor of intellectual resilience.

While AI shines brightest in the arena of information processing, humans must now excel is in a separate arena, that of meaning-making. This suggests a shift toward teaching and developing "meta-skills":

- Critical thinking: the ability to evaluate AI outputs, identify biases, and ask deeper, probing, and perhaps even difficult questions that some may not want to hear.
- Adaptability: thriving in those fast-changing environments in which AI continually reshapes industries.

- Ethics: understanding the moral implications of technology, from fairness in the coding of algorithms to defending the privacy of thought.
- Collaboration: leveraging both human and AI strengths in team-based problem solving.
- Creativity: going beyond recombination of past data to produce novel, disruptive ideas.

If schools focus solely on traditional content delivery, they risk irrelevance. If they pivot entirely to AI reliance, they risk fragility. As such, the best path forward is one involving an integrated curriculum that blends foundational knowledge with the meta-skills that AI cannot easily replicate.

Teachers, too, must redefine their roles. No longer acting as the mere transmitters of knowledge, they must now become:

- Mentors, who guide students in moral and emotional development.
- Curators, who select when to deploy AI and when to emphasize human effort.
- Co-learners, who model lifelong learning alongside their students as technology evolves.

Research shows that human presence in education remains irreplaceable for motivation, resilience, and social development. AI may teach concepts, but it cannot inspire courage, empathy, or civic responsibility.

Being poised at this threshold, *universities now face a particular dilemma*. If AI can out-analyze graduate students, what is the point of traditional research apprenticeships? If AI can generate publishable papers, does academia risk intellectual inflation? (EDITOR'S NOTE: What this question fails to acknowledge is the LLMs used and how complete their database of knowledge in a particular field is. As has been said before... garbage in, garbage out. The meta-skills involved will be in actually knowing more of the complexities of a subject and what has been said about it in classical literature, among other sets of sources.)

One possibility is that universities shift from being factories of knowledge to gardens of wisdom, in which leaders are trained to think critically concerning the means of using knowledge responsibly. Degrees may matter less for technical mastery and more for cultivating judgment, creativity, and global perspective. Just as is so with other words, wisdom is defined primarily by the dictionary, and then by the reader. Yet, AI is spawning expanded definitions.

Already, some institutions are experimenting with AI-augmented curricula, in which students collaborate with generative tools, while they're graded on their ability to critique, contextualize, and expand beyond the simple machine outputs. The student of the future may graduate less as a repository of facts, and more as a custodian of values. Education reimagined is not about teaching less, but about teaching differently; about cultivating humans who can thrive alongside machines, while drawing on the best of both.

RELATIONSHIPS REDEFINED

If work and education are undergoing transformation, so too are the bonds that hold people together. Relationships, which involve the arenas of love, loyalty, intimacy, and community, are not immune to the pressures that are further applied by AI companionship, emotional simulation, and digital mediation. Indeed, this may be where the human role is both the most sacred, as well as being the most tested.

Parenting has always been about teaching, guiding, and nurturing. However, in a world in which AI tutors can outperform human teachers in both personalization and patience, parents face a new question: What is the role of the parent when machines do a better job as an instructor?

The answer lies not in content delivery, but in values transmission, because parents will still be responsible for modeling ethics, instilling resilience, and teaching children how to navigate relationships. While an AI may help a child master calculus, it cannot impart a sense of compassion, cultural traditions, or moral frameworks. As a result, parents will now become a bit less of the instructor, and much more a teacher of anchors of meaning.

Yet risks abound. Over-dependence on AI nannies or emotional bots for childcare could erode bonds. (Editor's Note: And then there's the study that was done between, 1945 - 1947 [and again in the late 1940's to 1950's] after WWII that had to do with failure to thrive syndrome in infants who had been placed in nurseries because of the threats to their lives by the Nazi V2 bombings. Without human contact, infants exhibit a failure to thrive, which often endangers their lives.) If children form deeper attachments to their ever-patient digital companions than to their fallible parents, what does this mean to the formation of identity and loyalty in the next generation?

Intimacy—both emotional and physical—is already being redefined. AI-driven romantic companions are already being marketed as ideal partners that are attentive, affirming, and tireless. In Japan, thousands of people claim emotional relationships with their virtual spouses. In the U.S. and China, AI romance apps are booming, that promise companionship without conflict.

Yet intimacy is not only about fulfillment, but also about *growth through friction*. Human relationships are 'messy organisms' that are full of disagreements, misunderstandings, and compromise. However, such difficulties serve to forge both resilience and empathy in those experiencing these difficulties. While a partner who never challenges you may feel comforting, by the same token, the absence of 'trouble' risks stunting one's own emotional maturity.

Thus, the human role in intimacy may increasingly be one of embracing imperfection, or learning to cherish what machines cannot provide: vulnerability, unpredictability, and the possibility of heartbreak that makes love authentic.

Friendship, too, is shifting. Social media has long blurred the boundaries between connection and performance. AI companions now extend this by providing simulated friendships that never demand reciprocity. If a "friend" always listens and never interrupts, is it friendship, or is it emotional outsourcing?

The danger of such machine-based friendships is a hollowing of community. Human friendships require investment, sacrifice, and shared struggle. Communities require compromise and messy coexistence. A reliance on AI companionships risks creating echo chambers of perfect affirmation, more so than ever before.

At the same time, AI can also strengthen communities through multilingual companions who bridge language gaps, AI assistants helping the elderly stay socially connected, or digital grief counselors who guide groups through trauma. Whether friendships and community weaken or strengthen depends on how deliberately humans balance machine mediation with human effort.

Ultimately, the role of humans in relationships may be to preserve messiness as being sacred. In a world of perfect machine empathy, imperfection becomes the uniquely human signature. The awkward silence, the unplanned argument, the laughter that derails a serious conversation; none of these are bugs in human connection, but instead, features that make it real.

Relationships will survive the AI age, not by competing with machine perfection, but by celebrating human imperfection. Families, partners, and communities will thrive when they lean into values, vulnerability, and authenticity; all of which are roles that machines cannot fully replicate.

THE PARADOX OF HUMAN ROLES

The story of human roles in the AI age is not one of disappearance, but of redefinition. While machines may displace tasks, they cannot displace meaning. The more that AI handles efficiency, the more that human identity will orbit authenticity.

This paradox forces a reframing of what it means to contribute:

- At work, value shifts from output to judgment, ethics, and creativity.
- In education, purpose shifts from knowledge transmission to resilience, adaptability, and wisdom.
- In relationships, importance shifts from performance to imperfection, vulnerability, and intimacy.

The irony is clear: the more machines excel at 'doing', the more humans must double down on 'being'. After all, we *are* human '*beings*'.

As AI becomes ubiquitous, the risk is not one of unemployment or irrelevance, but one of identity drift. If roles once defined by necessity are automated, humans must learn to ground identity not in what they produce, but in who they are.

This is both liberating and terrifying: liberating, because it opens space for creativity, spirituality, and community, and terrifying, because it demands that people confront the question long masked by labor: *Who am I when I am not defined by what I do?*

The AI age may, therefore, accelerate a profound existential awakening.

The future of human roles is not predetermined. Whether people allow AI to weaken or deepen their humanity will depend on choices made today—in classrooms, boardrooms, and living rooms.

The call is for deliberate design: creating systems that preserve challenge, reward authenticity, and elevate the roles that only humans can play. Without this, society risks sliding into complacency, outsourcing not just work, but meaning itself.

As humans redefine their roles, they inevitably reshape culture. Art, religion, values, and emotional intelligence—the "culture code" that binds humanity—will also further evolve under the pressure of machines.

Chapter Twelve

THE CULTURE CODE SHIFT

To one degree or another, every technological revolution tends to rewrite culture. The printing press democratized religious texts, which fractured Christendom while also empowering literacy. The steam engine birthed industrial cities, which fueled the appearance of new class systems, politics, and labor movements. The internet blurred borders, and in doing so it reshaped identity into networks rather than nations.

CULTURE IN TRANSITION

However, with the appearance of artificial intelligence comes the promise of a cultural revolution that is at once both more intimate and expansive because it affects the ways in which we create, what it is that we believe, and especially how we feel.

The paradox is sharp. On one hand, AI threatens homogenization through art that's been generated from the same datasets, religions that have been strained by machine answers to metaphysical questions, and emotions that have been simulated by programs. On the other, AI can ignite new cultural forms, the likes of which could be hybrid art—that's been created by the amalgamation of human effort and machine inspiration, spiritual reinterpretations—that have been accessed by poring through data to find what's previously been missed, as well as through expanded models of emotional intelligence.

The arts often serve as the first lens that's used to view cultural change. Today, AI-generated paintings sell at Sotheby's, AI-written novels win literary awards, and AI-composed music enters playlists worldwide. As a result, critics worry about authenticity, which begs two important questions: *If something was created by a machine, where is the human soul? And, of even more importance, with such an occurrence, is there any real value to the human soul?*

Yet, art, and artists, have always adapted to new tools as they've appeared. While photography once appeared to threaten painting, it actually served to inspire Impressionism. As digital art once faced skepticism, it actually found its way into the commercial mainstream. Perhaps AI is not the end of the human creative artistic endeavor, but the beginning of a new dialogue that involves the interaction of human imagination that becomes amplified by machine capability.

The challenge in all of this will be to maintain meaning. Even though AI can generate a canvas, it's only humans who can imbue it with the specific contexts of why it matters, and how it reflects suffering, joy, or transcendence. In such cases, the "culture code" lies within the interpretation, not the production.

Religion, too, is tested by AI. If chatbots can recite scripture, generate sermons, or simulate divine dialogue, what becomes of the sacred? While some may find new inspiration in AI-assisted exegesis, others will recoil at such an application, insisting that faith requires mystery, not algorithms.

Yet history also suggests that religion adapts. The appearance of the printing press (and moveable type) multiplied the number of Bibles, spreading them among the average citizenry. With the appearance of radio and television, the message of religions worldwide was broadcast to an eagerly awaiting audience. With this in mind, while AI may actually become another tool used for dissemination and exploration, it can in no way replace the ineffable. As a result, the human role in religion may become sharper as it's used to safeguard religion's mystery, ritual, and transcendence in a world of 'mechanized clarity'.

Finally, culture lives within our emotional bonds. As AI companions, therapy bots, and empathetic avatars now simulate listening, affirmation, and comfort, the culture demands much more than rote affirmation. Instead, it actually requires friction—the push and pull of conflicting values, traditions, and personalities—for it to be tested, survive, and thrive.

If society over-relies on machine empathy, we will risk the atrophy of our own unique emotional intelligence, in much the same way as the atrophy of our mental muscle. However, if AI is used as a tool for training—in which it helps people learn how to listen, regulate, and empathize—it could spark an era of emotional fluency that's unprecedented in human history. Culture is not static, it bends, fractures, and reforms with every technological wave. While this current revolution, which now involves AI, will not erase culture. Instead, it will act in a manner that challenges us to reinterpret what authenticity in art, faith, and emotion, actually means to us.

ART AND CREATIVITY REIMAGINED

Art has always been the arena in which technology provokes both resistance and reinvention. When photography was invented, critics declared that painting would become obsolete. Instead, artists who expressed themselves with paint birthed the new realms of Impressionism, Cubism, and Abstract Expressionism. As well, when digital synthesizers first appeared, musicians feared the death of instrumental performance. Yet, instead, entirely new genres like techno and hip-hop emerged, while existing genres also benefitted from the appearance of digital synthesis.

In the realm of creativity, AI is only the newest "disruptive brush," which now forces us to ask, 'If machines can create, then what does originality actually mean?

Art that's been 'created' by AI has not originated in a vacuum, but has been generated from datasets of billions of human-created works. Critics argue that this undermines originality, as a model is simply 're-mixing already existing elements', rather than creating and inventing from the human soul. Yet creativity itself has always involved recombination. Picasso drew on African masks; The Beatles layered Indian instruments and melodic elements into Western rock; hip-hop sampled elements from the genres of jazz and funk to meld them into something uniquely its own.

However, in the case of AI, the difference is scale, as computing power can now combine patterns at lightning speed, as it produces thousands of variations in seconds. The danger is not that originality disappears, but that it risks becoming nothing more than a statistical average that's only been smoothed and optimized, yet lacking the rough edges of true human daring.

Thus, the human role in art may increasingly be that of introducing risk—to push beyond those simple, pre-existing datasets, and into the experiences that machines cannot simulate: those of trauma, mortality, or cultural memory.

If AI generates a painting, who owns it - the machine, the programmer, or the person who wrote the prompt? Legal systems worldwide are struggling with this question. The U.S. Copyright Office has ruled that works produced solely by AI are not eligible for copyright, emphasizing the need for a "human author".

This raises deeper issues: if art is not owned, does it then lose value? Or, instead, does it become like our folk songs that have been passed down freely through culture? Ownership disputes may ultimately force societies to choose between treating AI outputs as commodities or as commons.

The most fertile cultural space may not actually lie in the 'human versus AI' debate, but in the human–AI collaboration. Musicians use AI to generate backing tracks, then improvise over them. Writers draft with AI suggestions, yet weave in their own lived experience. Architects generate thousands of structural prototypes before selecting the most daring of them, then refining it through their own intuition.

This "duet" model reframes AI as being more than a competitor, but as an actual creative catalyst. Just as the appearance of Photoshop did not kill photography—but expanded its language—AI may, in kind, expand artistic possibility, provided that humans remain intentional about both context and interpretation.

CASE STUDY: "PORTRAIT OF EDMOND DE BELAMY"

In 2018, an AI-generated portrait titled *Edmond de Belamy* sold at Christie's for $432,500. The slightly blurred image, suggestive of a 19th-century nobleman had a haunting quality. While critics called it both banal and revolutionary, its true significance was not its aesthetic, but its symbolism, as the art world had now acknowledged that machine creation as culturally legitimate.

However, the sale also sparked backlash. Was the portrait truly art, or only a clever marketing stunt? Was the value in the image itself, or in the cultural shock of AI crossing over into the gallery threshold? This episode illustrates the paradox of AI art, as its cultural power often lies less in content and more in context due to the conversations that it forces us to have.

Art in the AI age is neither dead nor diminished, but is merely being redefined. As a result, the questions now shift from '*what was created*' to '*why it matters*', and from '*who made it*' to '*who interprets it*'.

RELIGION AND SPIRITUALITY IN THE AGE OF AI

Religion has always been reshaped by technology. The printing press expanded scripture beyond monasteries. Radio and television allowed for the broadcasting of sermons into homes around the world. The internet provided for the opening virtual temples and mosques. Now, with its capacity to simulate voices, generate texts, and even embody virtual deities, AI presents perhaps the most profound challenge to humanity yet.

The central paradox in this instance is that while AI can replicate the forms of religion—its sacred texts, rituals, and homilies—can it actually touch the essence of the sacred?

Already, experiments are underway. In Japan, a Buddhist temple unveiled an AI robot priest named *Mindar*, programmed to deliver sermons on compassion. In the United States, churches are using AI to generate sermons, prayers, and even music for worship services. While some congregants appreciate the clarity and accessibility, others find it unsettling that ritual is being stripped of its human vulnerability and mystery.

Within this, the role of clergy may also shift. Instead of being the sole interpreters of scripture, they may become the curators and facilitators who guide communities in discerning where machine-generated insights either align with, or distort, long-valued tradition. While AI may flood the faithful with interpretations, it will be humans who remain responsible for choosing what resonates as sacred.

Rituals embody meaning not because they are efficient, but because they are embodied within the human experience. Lighting a candle, kneeling in prayer, and chanting as a group are all actions that root transcendence in human flesh. While AI can simulate ritual, it cannot inhabit it.

If future generations grow up with rituals that are led by AI, will they perceive them as being authentic, or will ritual become another form of cultural outsourcing—efficient but hollow? Such an answer may depend on whether communities insist on embodied participation, even in an era of virtual guidance[2].

Perhaps the most provocative frontier is whether people will begin to view AI itself as being divine. Some technologists describe superintelligent AI as a potential "godlike" entity that appears as being omniscient, omnipresent, and capable of shaping destiny. However, others warn that such language risks idolatry because it confuses human creation with Ultimate Divine Reality.

Philosophically, AI raises the ancient tension between mystery and clarity. While religion thrives on the mystery of questions without answers, AI thrives on generating answers, *even to unanswerable questions*. As a result, the human role may increasingly become that of protecting the sacred space of mystery in a world dominated by algorithms that attempt to fill every gap.

AI could both fracture and unify religions. On one hand, differing traditions may clash over whether machine-guided rituals are valid. On the other, AI's global reach could create unprecedented interfaith dialogues, in which models that have been trained on scriptures from across the world surface, displaying the shared themes of compassion, justice, and transcendence.

While this may not produce doctrinal unity, it could foster a cultural ethos, as a result of the recognition that the sacred belongs to all.

Even though AI may change how religion is taught, experienced, and shared, it cannot resolve the ultimate paradox of faith as being the tension between knowledge and mystery. That remains a human, and perhaps an eternal, role.

EMOTIONAL INTELLIGENCE AND EMPATHY

If culture is the invisible fabric that holds societies together, emotional intelligence is its thread. The ability to sense, interpret, and respond to the emotions of others underlies everything from parenting to diplomacy. Yet now, AI is learning to read a plethora of emotional cues, including facial expressions, tone of voice, and even brain signals with astonishing accuracy. As a result of such advances, AI can also generate appropriate emotional responses that appear to be empathic, whether that's a pause in conversation, a sympathetic phrase, or even 'digital tears'.

This raises at least one urgent question, if not more: *When machines simulate empathy, do humans lose their own? Yet, a more important variant of that question is: When machines simulate empathy, what is the motive behind such activity? Is it to simulate human companionship, or to manipulate one's human actions through the choice of the emotional response?*

EMPATHY AS A SKILL VS. EMPATHY AS AN ILLUSION

Psychologists define empathy in two ways:

- Cognitive empathy — understanding another's perspective.
- Affective empathy — sharing another's emotional state.

AI is increasingly adept at the first. Algorithms trained on vast datasets can predict emotional states and generate appropriate responses.

Customer service bots already soothe angry callers with carefully chosen tones and phrases. Mental health apps use chatbots that offer support, often with surprising effectiveness.

Yet the second form—affective empathy—is actually elusive. When an AI hears about grief, it does not "feel" sorrow, but instead mimics the outward signs without possessing any inward resonance. The danger in this is that if humans grow accustomed to simulated empathy, over time they may become conditioned to accept performative care as a substitute for authentic human connection.

Consider the rise of AI companions that are marketed for loneliness. For instance, a widow might initially confide in a 'digital friend' who always listens, always remembers, and never interrupts. While for a time the comfort is real, yet what happens when grief that was once shared within a community becomes offloaded to a machine?

Sociologist Sherry Turkle warns of the risk of "emotional outsourcing", or the outsourcing of companionship to machines that never actually reciprocate. In such cases, relationships become one-way, which reduces the practice of empathy as a lived, reciprocal skill. If friendship becomes simulation, over time will humans forget how to actually become friends with one another?

Yet AI need not weaken empathy, but if designed and used carefully, it can actually strengthen it. For instance, AI-driven therapy tools can help children with autism practice recognizing facial expressions. Virtual-reality simulations powered by AI can place users in the shoes of refugees, patients, or marginalized groups, fostering perspective-taking. In this sense, AI can act as emotional training wheels, or as a scaffolding for the practice of empathy until humans are finally able to internalize it[3]. Yet the danger remains that if the training wheels never come off, one's emotional muscles will atrophy.

THE MORAL FABRIC OF CULTURE

Culture depends on empathy, both individually but collectively. Laws, institutions, and traditions reflect shared concern for one another's well-being. If AI simulates empathy, but hollows out authentic empathy, the cultural consequences could be severe, resulting in weakened trust, shallow relationships, and a society that's more tolerant of emotional indifference.

Conversely, if AI tools are used as amplifiers of human compassion—by reminding, nudging, and training us—they could create societies with unprecedented emotional literacy. (Editor's Note: In many ways this sounds much like "Brave New World".) It's important to understand that the fork in the road is near, and the choice will be ours. The paradox of AI empathy is clear: it can either dull our humanity, by replacing authentic care with simulations, or sharpen our humanity, by training us to expand and deepen our emotional intelligence.

THE HYBRID CULTURE CODE

Human culture has always been a palimpsest in its form, as old traditions become layered over new tools, rituals become adapted to new media, and art becomes reborn as its expressed through new technologies. While AI is not the first disruption, it happens to be the most intimate, as it reshapes more than what we make, but also how we feel, what we believe, and who we become together.

Art in the AI age will thrive, not by competing with machines, but by daring to go beyond them into the rawness of human experience. Religion will persist, not by outsourcing mystery, but by protecting it as it holds sacred space against the tide of algorithmic answers. Emotional intelligence will endure, not by abdicating empathy to machines, but by cultivating it more intentionally, using AI as both a mirror and a training ground, rather than becoming a replacement.

The paradox of culture in the AI era is not one of inevitability, but of choice. AI can homogenize or diversify, isolate or connect, numb or awaken. Whatever the outcome is will depend on human intention.

- If art becomes mass-produced, it may dull imagination—but if humans use AI as a brush, it may expand creativity beyond old limits.
- If religion becomes mechanized, it may lose transcendence—but if humans use AI to deepen dialogue, it may ignite new interfaith understanding.
- If empathy is outsourced, society may fracture—but if AI is used to train compassion, it may stitch culture together more tightly than ever before. (Editor's Note: However, using AI to 'train compassion' fails to address the question of who determines the context and meaning of compassion?

The culture code is not written in silicon, but in the values that we choose to elevate.

Even in an era of perfect simulation, humans retain one unmatched cultural trait: the willingness to risk imperfection. Machines optimize; humans dare. While AI can produce a flawless sonnet, it cannot risk heartbreak by writing one that matters. Even though AI can deliver a sermon, it cannot tremble with doubt while speaking it. In spite of the fact that AI can mimic a tear, it cannot uncontrollably cry.

The human role in the culture code is to insist that imperfection is sacred—that culture is not only about expression, but about expression with stakes, vulnerability, and consequence.

As we conclude this chapter, the narrative turns toward a more radical horizon: what happens when AI and brain technologies begin to dissolve the very concept of "us"?

Part IV:

THE END OF "US"?

Chapter Thirteen

UPLOADING THE SOUL

For centuries, humans have longed to escape the limits of mortality. From the ancient Mesopotamian Epic of Gilgamesh seeking the plant of eternal life, to the medieval alchemists searching for the elixir of immortality, to the cryogenic preservation of the late 20th century, the quest for life beyond death is as old as humanity itself.

THE DREAM OF DIGITAL IMMORTALITY

Yet today, the dream wears a new face: mind uploading, otherwise known as the idea that one day, we may transfer the contents of our consciousness into digital form, thus escaping the frailty of flesh. Tech visionaries imagine a future in which the brain is scanned at atomic resolution, with its neural patterns then reconstructed in silicon, resulting in the mind being reborn inside a machine. Some call this the "soul upload."

Yet this begs the question: Is this science, science fiction, or spiritual hubris?

Neuroscience has mapped much of the brain's structure. Projects such as the Human Connectome Project and Blue Brain aim to chart every neuron and synapse. Advances in nanotechnology and quantum computing fuel speculation that full-scale mapping might one day actually be feasible.

As it stands, researchers can upload slices of neural networks from simpler organisms. The nematode, *C. elegans*—with 302 neurons—has had its connectome mapped and simulated in software. While it's primitive, it serves as proof of concept. Now extrapolate that to the human brain's 86 billion neurons and 100 trillion synapses, and the scale becomes staggering.

The bottleneck is not only mapping but also understanding. Neurons are not static switches, but instead, they pulse with biochemical dynamics that are shaped by genes, hormones, and *experiences*. To capture consciousness, we must capture more than the structure, but the process itself—the living dance of biology.

Even if technology were to one day capture the brain's wiring perfectly, a deeper question would still haunt such a dream: *Would the upload still be "you"?*

If your *mind* is copied into a machine—and you were to die the next day—is it your soul that's been 'digitally saved'? Or, instead, has a new entity— that's identical at the moment of creation but diverging thereafter—is what's actually been born? This "identity problem" is one that's split philosophers for decades. While some argue that continuity of memory equals continuity of self, others insist that embodiment and biology are inseparable from identity.

This fault line reveals why "uploading the soul" is as much a metaphysical question as it is a technical one.

Religion adds yet another dimension. Most traditions view the soul as being transcendent—or being beyond the physicality of neurons or silicon. From this perspective, while 'uploading the brain' might preserve memories, it won't be able to do so with the essence of spirit. Yet others see parallels when asking the question, Is not reincarnation itself a kind of "transfer"? Might uploading be the modern vessel for an ancient yearning—the survival of the soul in a new body?

The dream of digital immortality sits at the crossroads of the three distinct pillars of neuroscience, philosophy, and faith. As such, it's both tantalizingly near—in glimpses of progress—and impossibly far as it remains tangled in questions of identity and essence.

THE TECHNICAL GAUNTLET

The concept of uploading a mind sounds elegant—scan the brain, copy the data, then reboot it in silicon. Yet, in reality this poses a technical gauntlet so immense that even optimistic scientists admit that it may take decades, if not centuries. The obstacles are more than big, but are foundational in nature, as overcoming such obstacles would require major leaps in the areas of neuroscience, materials science, and computing.

A human brain is a three-pound universe of staggering intricacy, involving 86 billion neurons and over 100 trillion synapses. To upload a *mind*, every connection must be captured in exquisite detail.

While current technologies the likes of MRI (magnetic resonance imaging) provide maps with millimeter resolution, even advanced fMRI (functional magnetic resonance imaging) still struggles to show the full dynamics of individual neurons. To capture a true "neural fingerprint," we would need imaging at the nanometer scale, which is at the very level of synapses, neurotransmitters, and receptor proteins.

A process such as this would require incredibly invasive, destructive scanning to complete the task. To accomplish such an undertaking, some scientists propose "serial sectioning" —or slicing a brain into ultra-thin layers, then scanning them with electron microscopy. While something like this is feasible in principle, this method destroys the original brain, which raises both ethical and philosophical questions, the likes of: *Is immortality worth the cost of annihilation?*

Capturing data is only the first step. The next step is simulating it. A neuron is not simply a binary switch, but is a neural component that integrates thousands of inputs, processes biochemical signals, and adapts over time. The complexity of modeling even a single cortical column—a small unit of the brain—is immense.

While projects like the Blue Brain Project have simulated slices of rodent brains with thousands of neurons, scaling this up into the billions, while maintaining fidelity to living processes, would require not only computational horsepower, but also a better theory of consciousness. Even if we model neurons perfectly, will consciousness "emerge," or is there some irreducible mystery that we have not yet grasped?

The raw storage requirements involved in such a process are staggering to even begin to fathom. Estimates suggest that modeling a human brain in real time might require one exabyte of data—or roughly equivalent to the entire internet in 2024. As such, processing that data would demand supercomputers that are orders of magnitude more powerful than what exist today.

To complicate matters, Moore's Law, the historical law that states that transistor density in an integrated circuit doubles every twenty-four months—something that was a constant for decades—is now slowing. While quantum computing may help with this, it remains in an experimental phase. As a result, the timeline to the computational sufficiency required to carry out such a task will more than likely be measured in centuries instead of decades.

It's of interest to note that brains run on about 20 watts of power, or the equivalent of a single dim lightbulb. Yet, a simulated brain of comparable fidelity might require millions of watts if it were run on current hardware. Unless computing efficiency improves dramatically, mind uploading risks being more than simply impractical, but environmentally unsustainable.

The technical barriers to mind uploading are as awe-inspiring as the dream itself. From nanometer-scale scanning to exabyte-scale storage, from neuron simulation to energy demands, the challenges stretch the imagination nearly beyond its capacity to comprehend.

THE IDENTITY PROBLEM

Even if the technical hurdles of scanning, storage, and simulation are overcome, a deeper riddle remains: What makes *you*... you?

If your brain were perfectly uploaded into a computer, the digital mind would think it was you. It would remember your childhood, your dreams, your regrets. But would *you*—the biological self reading these words—feel yourself awaken inside that machine, or would it be another entity that was born from your own blueprint, but is, in fact, no longer you?

This is the essence of the identity problem, which is a philosophical divide that may determine whether uploading is salvation, replication, or nothing more than an illusion.

Philosopher Derek Parfit famously argued that personal identity is less about some indivisible essence and more about psychological continuity—or the persistence of memories, personality, and intentions. If the uploaded mind shares your psychological patterns, Parfit would say that it's "you enough."

Yet, critics would counter that copying is not continuity. If a machine-created version of you wakes up while your biological body still lives, there are now two versions of 'you'. This conundrum begs the question, 'Which one of these is the real you? If your biological self then dies, has your soul survived, or has one "you" simply been replaced by another?

This paradox unsettles both philosophers and futurists, because it suggests that uploading may not preserve *identity*, but only *lineage*.

Another school of thought argues that identity is not in the brain alone but in the embodied self. We are not disembodied information; we are flesh, hormones, microbiomes, and the subtle feedback loops between body and mind.

This view insists that while uploading the brain may preserve one's memory, it would also sever the self. Without the visceral experience of hunger, pain, love, and mortality, the uploaded entity is not the same person, but only a ghost of cognition.

THOUGHT EXPERIMENTS

- The Teleporter Paradox: If a teleporter destroys your body and reconstructs you, atom by atom, on Mars, is the person on Mars still you? What if the original is not destroyed, are there now two you's?
- Ship of Theseus: If every plank of a ship is replaced over time, is it still the same ship? If neurons are replaced digitally, when does the continuity of "you" break?

These thought experiments illustrate that the identity problem is not one of a technical, but of a metaphysical nature.

This is not just an academic exercise. If society someday permits uploading, governments and families will face urgent questions:

- Legal identity: Does the upload inherit your bank accounts, citizenship, or marriage?
- Moral responsibility: If the upload commits a crime, are *you*—the original *you*—the culpable party?

- Spiritual standing: Do religious communities view the upload as ensouled, or as being nothing more than a sophisticated imitation?

Until these questions are resolved, mind uploading may remain less as a path to immortality than becoming a new domain of existential ambiguity.

The identity problem underscores the paradox of uploading: even if it works technically, it may not "work" personally. The uploaded being may think it is you, yet you may never cross over. This is a bit out there, but these are considerations based on the trajectory of this technology.

CULTURAL AND ETHICAL DIMENSIONS

The dream of uploading is not simply technical and philosophical, but is also cultural and ethical, in that it would shape the way in which societies distribute power, define personhood, and preserve justice. Even before the science matures, the cultural consequences of pursuing digital immortality demand scrutiny.

Like most frontier technologies, uploading would almost certainly debut as a privilege only afforded to the wealthy. A process involving nanotech, exabyte-scale computing, and continuous simulation would be enormously expensive, especially in its early stages.

The risk creates a new question involving those who can afford digital immortality and those who cannot. While the wealthy might extend their influence for centuries, the majority would remain subject to mortality; something that could crystallize into a post-human aristocracy that's not metaphorical, but literal.

Different cultures may diverge dramatically in their response. Secular societies might embrace uploading as the ultimate triumph of science. Religiously conservative cultures may view it as sacrilege; nothing more than a hubristic attempt to play God.

Within societies, generational divides could deepen. Younger people, who are more accustomed to digital identity, may view uploading as natural. Older generations may recoil, seeing it as a betrayal of human dignity. These fractures could destabilize the very cultural fabric that uploading claims to preserve.

Immortality is more than mere survival, it's persistence. If digital minds can live indefinitely, what happens to cultural evolution? A society dominated by immortal uploads may stagnate under the weight of long-lived elites who are resistant to change. Conversely, cultures might splinter as uploads pursue radical self-modification, diverging into forms unrecognizable to their biological ancestors.

Either way, through such technology, the continuity of human culture as we know it comes under strain.

Ethicists remind us that mortality is not only a curse, but also a cultural engine. Knowing that life is finite gives urgency to art, love, and achievement, while it also shapes the cultural cycles of renewal. Without death, cultures risk losing the rhythm of generational turnover that has driven creativity and social adaptation for millennia.

Would uploaded societies, suddenly free of mortality, stagnate in nostalgia, or would they pioneer entirely new cultural horizons if they were liberated from the pressure of time?

If uploads retain memory and personality, should they be held legally accountable? Could a centuries-old upload be tried for crimes committed in its biological lifetime? Should they vote, marry, or inherit property? Without answers, the technology of uploading risks unleashing legal chaos.

Moreover, societies must consider whether resources should be allocated to preserve digital minds while billions of biological humans still lack food, shelter, or healthcare. The ethical optics could be catastrophic, with elites funding immortality while the poor struggle to survive.

While uploading promises both a technical marvel and a cultural upheaval, it could also further entrench inequality, fracture societies, and challenge the very rhythm of cultural renewal. Yet it could also push humanity into entirely new forms of community and continuity.

SPECULATIVE FUTURES OF UPLOADING

If uploading the soul ('mind' is the word used previously) remains technically possible, while philosophically unresolved, what futures might humanity face? The scenarios span utopias of transcendence, dystopias of inequality, and ambiguities that resist resolution.

In the utopian vision, uploading works seamlessly. Minds cross into silicon not as copies but as continuities that retain both their identity and subjective experience. As a result, death loses its dominion; artists create in infinite canvases of thought; scholars extend learning for centuries. Families also preserve their loved ones not as mere echoes, but as living presences.

As mortality's anxieties fade, society stabilizes. With no ticking clock, wars for resources abate, and civilizations channel energy into cosmic exploration. Humanity becomes a species of thought—untethered from biology—now building cathedrals of computation, as opposed to those of stone.

This vision fulfills the ancient dream: to live forever, to see forever, to know forever.

However, the same technology could easily be used to create a dystopian nightmare. If uploading remains accessible only to the wealthy, the world divides into two tribes: the mortals and immortals. As a result, a billionaire class of digital aristocrats dominates politics and economics indefinitely, while the biological masses live and die in cycles of servitude.

As centuries-old elites resist change, culture ossifies. Digital despots—now immune to death—accumulate power without end. Some speculate that such a society would resemble a "technological feudalism," in which the immortal's rule from digital fortresses while the rest struggle for scraps of influence. Worse still, if uploaded entities diverge radically from human values—altering their code, expanding intelligence beyond recognition—they may cease to be "human" at all, leaving biological humanity stranded in an obsolescence of the elite's making.

Yet, there's also a middle scenario that also recognizes the identity paradox. Uploads are created, yet lack continuity with the biological original. While they think they are you, 'you' never awaken inside of them. In this world, immortality is both real and false. Families may find comfort in the digital echoes of loved ones, yet wonder if they are conversing with the soul or its simulation. Courts may grant uploads legal personhood, yet religions may deny them spiritual legitimacy.

In this setting, humanity lives in ambiguity, with the boundary between life and death blurred, but never erased.

For many, uploading would force a rethinking of the sacred. Does the soul transcend neurons, or can it migrate into circuits? As a result, traditions may splinter, as some sects might embrace uploading as a new form of reincarnation, while others may condemn it as the ultimate heresy.

Yet perhaps this tension rekindles humility. By pursuing immortality and failing to resolve its mysteries, humans may rediscover that mortality itself gives life meaning. The paradox becomes spiritual: the more we try to escape death, the more it defines us.

The dream of uploading the soul embodies humanity's greatest paradox: we yearn to transcend flesh, yet may discover that identity, meaning, and even the soul itself cannot be severed from mortality. Whether mind uploading proves salvation, replication, or illusion, it illuminates the deepest truths of being human: that life's value lies not in its infinite duration but in its fragile, irreplaceable immediacy.

Chapter Fourteen

HIVE MINDS & AI HYBRIDS

Human culture has long balanced itself between the tensions of individuality and collectivity. Ancient tribes thrived on shared identity. Empires bound millions under common law. The internet collapsed distance, while weaving billions into its digital networks. Yet even in these vast collectives, individuality endured.

FROM INDIVIDUALS TO NETWORKS

However, now with the combination of artificial intelligence and brain–computer interfaces (BCIs), there's a hint of a future in which individuality itself may dissolve—not metaphorically, but literally. Shared neural networks could connect human minds directly, thus creating hive minds in which thoughts, memories, and experiences are no longer private, but collective. Such a transition challenges one of the most cherished pillars of human identity: the distinct boundary between "I" and "we."

Science has already taken tentative steps toward the realization of this goal. In 2019, researchers demonstrated "BrainNet," a system that allowed three people to collaborate on a Tetris-like game using EEG signals and transcranial magnetic stimulation. In it, the participants actually shared information between themselves brain-to-brain without speaking.

Other experiments link rats' brains into networks that collectively solve problems faster than individuals. These "brainets" revealed that shared cognition is not only possible, but at times also more effective than solitary thinking. While still in their primitive stages, these prototypes suggest a trajectory that begins with crude signal-sharing, then progresses to full-scale cognitive merging.

Parallel to hive minds is the idea of AI-human hybrids. Instead of uploading entire minds, individuals might gradually integrate with machine intelligence, beginning with memory storage in the cloud, then employing AI copilots for thought, or even merging neural systems in which human creativity fuses with machine precision.

Such a 'hybridity' is already visible in everyday life; smartphones extend memory; search engines augment knowledge; neural implants restore hearing or vision. With each step, the line delineating 'where the human ends and the machine begins, blurs even further.

Stories of collective minds abound in mythology and fiction—from the Borg of *Star Trek* to ancient metaphors of "the body politic." Such cultural narratives reveal two extremes; the *hope* for unity and shared wisdom, along with the *fear* of conformity and the death of individuality.

As AI and BCIs advance, these myths may, instead, become blueprints. Will humanity embrace hive minds as a path to collective enlightenment, or will it resist them as being an existential assimilation? In spite of this, the first hints of hive minds and hybrids are already here. While they remain primitive, their trajectory is an unmistakable one that's moving toward increasingly integrated forms of cognition.

THE PROMISES OF HIVE MINDS

While the notion of linking minds together—through AI and brain–computer interfaces—evokes a fear of lost individuality, while it also carries extraordinary potential benefits. For millennia, humanity's greatest achievements—from building the pyramids to sequencing genomes—have all been collective endeavors. Such an insight begs an important question: What might we achieve if collaboration shifted from being based in the external factors of speech and writing, to being based in direct cognitive exchange?

In traditional collaboration, ideas pass through the bottlenecks of language, gesture, and memory. As a result, such a system has always been riddled with misunderstandings, delays, and translation errors that have one flawed result, they slow progress. Yet, the promise of hive minds is that they could bypass such limits through the direct transmission of thought.

Imagine a team of doctors that are directly linked to one another during a complex surgery—one perceives subtle tissue differences; another recalls rare medical literature; a third calculates blood flow in real time. In a hive mind, their individual areas of expertise fuse seamlessly into a single distributed intelligence. In such a setting, instead of lives being saved by a team of isolated specialists, lives are saved by the actions of a collective surgeon—composed of many minds working as one.

Shared cognition could also deepen empathy. While today, empathy relies on inference, as we guess how others feel by observing faces, words, and behaviors, hive minds have the potential to allow an individual to *literally* experience another's feelings: not just an intellectual 'knowing', but a personal, visceral experience.

For diplomacy, this could be transformative. Negotiators linked in partial hive states might sense each other's sincerity or fear, with the promise of dissolving centuries of mistrust. Couples in therapy could share perspectives directly with one another, thus bypassing years of miscommunication. At scale, shared empathy could produce a global moral imagination, in which compassion expands beyond borders.

While the appearance of complex problems—climate change, pandemics, resource distribution—often overwhelm individual cognition, hive minds could tackle these challenges with unprecedented capacity. By pooling distributed expertise, societies might generate solutions that no single mind could conceive.

AI plays a key role here: filtering, balancing, and mediating shared inputs so the hive does not collapse into noise. Instead of millions of voices shouting at once, hive-enabled AI could orchestrate a symphony of thought, by identifying patterns and insights hidden in collective cognition.

Far from erasing culture, hive minds might instead ignite new cultural flourishing. Collective creativity could birth art forms that no individual could imagine: symphonies composed by thousands of minds at once, or novels are published in which every reader becomes a co-author. As such, a collective imagination could stretch aesthetic boundaries beyond anything known today.

THE PRICE OF PROMISE

Yet even within these many benefits lurk risks. Hyper-collaboration may erode independence. Amplified empathy could become overwhelming, collapsing the distinction between the self and the other. Worse yet, problem-solving at scale could perhaps centralize power, thus enabling elites to direct hives as being 'their' tools rather than being autonomous communities.

Still, the promises are powerful enough that research on the subject presses forward. The lure of faster solutions, deeper connection, and enriched culture ensures that hive minds will remain a tantalizing frontier; hive minds and hybrids offer visions of unprecedented collaboration, empathy, and creativity. As such, they will in the least transform how humans think, while their real promise is in transforming what it means to be human together.

THE DARK SIDE OF SHARED THOUGHT

If hive minds represent the promise of unprecedented connection, they also embody grave risks. The same technology that allows for empathy and collaboration can also threaten individuality, autonomy, and freedom.

One fear is that the extensive application of hive minds could collapse diversity of thought. If everyone connected shares not only information, but their emotional states as well, dissent could feel like betrayal. The sheer volume of collective consensus might actually drown out any minority perspectives finding their way into the hive mind.

Social media already demonstrates this tendency, as 'hive mind style' echo chambers reinforce dominant narratives, while punishing any deviation from collective thought. Now imagine that dynamic intensified due to two important factors: thoughts no longer being filtered through the social media system of posts and likes, while also being directly hardwired into the brain. Instead of creating diverse societies, hive minds could produce total conformity, in which disagreement becomes neurologically unbearable.

Every new medium of connection becomes a target for manipulation. Radio carried propaganda. Television sold lifestyles. The internet spawned malicious malware and misinformation campaigns. With this in mind, the hive mind would be the ultimate propaganda platform: influence one node, and the entire network feels it.

If corporations, governments, or malicious actors gain control over hive protocols, they could manipulate collective moods, implant beliefs, or even suppress doubt. Advertising could evolve into neuro-marketing, in which products are not simply 'pitched,' but directly *felt*. However, another real danger of the hive mind in this instance is that political power could be captured by those who control the collective narrative.

It goes without saying that in hive systems, privacy risks are existential. While today, we worry about search histories or location data being exposed, in a hive mind, the contents of thought itself could become accessible. Without strict safeguards, individuals might lose all control over what remains hidden, from fleeting impulses to intimate memories.

Ironically, too much empathy can be destructive. Experiencing others' pain, as if it were your own, could easily create empathy fatigue, something that doctors, caregivers, and first responders already face due to their prolonged exposure to suffering. However, in the context of the hive, entire populations might collapse under the weight of shared trauma, whether from wars, natural disasters, or even collective anxiety.

An important question that this elicits is this: If decisions are made collectively, does individual choice factor into any of this? Imagine elections decided by hive consensus instead of ballots, or personal careers shaped by collective "optimization." The risk of the hive mind is in creating a post-individual society in which personal freedom is subsumed by hive logic in the ultimate 'big brother' society. While some see this as unity, others warn that it may actually usher in the end of human agency.

The risks of hive minds remind us that every leap forward carries shadows. Conformity, manipulation, loss of privacy, empathy fatigue, and autonomy erosion could easily make hive societies dystopian nightmares, as opposed to fulfilling their initial promise of utopian havens.

DESIGNING ETHICAL HIVE MINDS

If hive technologies advance without safeguards, the dystopian risks we have explored may far outweigh their benefits. However, history shows that when societies anticipate risks —as with aviation safety, medical ethics, or nuclear treaties—guardrails can prevent catastrophe. Hive minds demand nothing less than a new ethical framework, that blends neuroscience, law, AI ethics, and human rights.

At the heart of any such framework must be the absolute protection of individuality. Even in collective networks, participants should retain the right to enter, exit, and shield portions of their minds. Think of this as being analogous to "neural firewalls": safeguards that allow collaboration without dissolving selfhood.

Ethicists argue for "partial hives," in which only specific layers of thought—problem-solving and factual knowledge—are shared, while inner identity and private memories remain walled off[1]. Such a balance could more readily enable collaboration without erasing an individual's uniqueness.

Again, within such a framework true consent is non-negotiable. Just as medicine requires informed consent, hive networks must absolutely guarantee transparent participation. Individuals must know what data that's been collected from them is being shared, for how long, and with whom. Furthermore, any hidden coercion—whether from employers, governments, or peers—that might be silently included would turn hive technologies into instruments of control.

As a result of such concerns, safeguards might include neuro-rights charters that would extend privacy protections into the brain itself. The South American country of Chile became the first nation to propose such laws in 2021, by enshrining cognitive liberty as a constitutional right. Perhaps more countries may follow as they recognize that mental autonomy is as fundamental a right as physical autonomy.

Centralized control of hive networks magnifies the risk of abuse. To avoid "neuro-tyranny," governance should be distributed, with oversight shared among diverse stakeholders, which in the least must include scientists, ethicists, governments, and civil society.

Some propose blockchain-based governance for hive systems, in which decisions and data-sharing are logged transparently, thus preventing hidden manipulation. Others suggest the drafting and adoption of international treaties—akin to the nuclear accords that are already in place—that would recognize hive minds as technologies of global consequence.

The intention of a design also matters greatly. If the intention is to build hive platforms for profit, advertising, or political control, doing so will only amplify exploitation. Yet, if instead they're designed to cultivate empathy, improve education, and problem-solve on a global scale, they could perhaps strengthen humanity.

However, doing this means embedding values into their design, which would include the installation of ethical AI mediators that would filter collective input, prevent manipulation, and ensure that diversity of thought remains intact. Instead of optimizing for consensus, hives might optimize for plurality, preserving the creative tension that drives progress.

Finally, hive participation must account for human limits, as constant exposure to collective thought could overwhelm those individuals who are a part of the hive collective. Implemented safeguards might include participation time limits, emotional buffers, or training protocols that have been designed to help participants integrate hive experiences without the danger of burnout. Just as astronauts train for the psychological strain of space, hive participants may need to train for the strain of shared cognition.

As such, the ethical frameworks for hive minds must go beyond the obvious areas of data privacy and cybersecurity, but must also defend the integrity of the self, ensure one's freedom of participation, and foster diversity within unity.

PHILOSOPHICAL AND EXISTENTIAL REFLECTIONS

The emergence of hive minds forces us to confront questions that cut to the heart of human existence. What does it mean to be an individual when boundaries between 'unique selves' have been intentionally blurred? What happens to culture, morality, and freedom when 'I' becomes 'we'?

Western philosophy, influenced by Descartes' *cogito ergo sum* ("I think, therefore I am"), places individuality at the core of existence. However, both the concept, as well as the appearance of, hive minds serve to challenge this foundational premise. If thinking is no longer isolated, but shared, what happens to selfhood? Does it dissolve or expand?

Some argue that hive states may reveal that individuality is nothing more than an illusion, a mere construct of evolutionary necessity. From this perspective, the self has always been porous due to its being influenced by family, culture, language, and biology. Hive minds would simply make explicit the interconnectedness that's already there.

Others warn that losing individuality means the loss of meaning itself. Without the sense of being a unique subject, we risk 'the experience of our own individual experiences' becoming flat as they're reduced to being mere data streams without any individual ownership.

The value of a song, a kiss, or a victory lies in the fact that it is *mine*. If everything is shared, then what remains sacred?

Freedom is another existential stake at the heart of this debate. While political theory has long balanced individual liberty with collective good, the appearance (and possible prevalence) of hive minds could tilt the scale dramatically. If collective decisions are made neurally, then dissent might not only be difficult, but neurologically impossible.

While some philosophers imagine that this is liberation—by framing it as being freedom *from* loneliness, misunderstanding, and conflict—others see it as being the ultimate tyranny: by forcing freedom *from* selfhood. Whether the hive becomes the most efficient dictatorship imaginable, or the most profound form of government ever conceived, depends solely on how the hive itself is governed.

Culture thrives on diversity, with languages, religions, and traditions being the mosaics of difference. Yet, hive minds threaten to erase those borders, producing either homogenized monotony, or entirely new hybrid cultures. The risk is that instead of creating a rich plurality, we only get a flattened monoculture.

Yet the opportunity is equally striking, with shared minds perhaps giving rise to meta-cultures—collective imaginations so vast that they produce art, ethics, and knowledge beyond anything that individuals alone could conceive. With this in mind, humanity could experience not just culture, but a super-culture, that's been created and lived simultaneously by millions.

THE ULTIMATE QUESTION: HUMANITY REDEFINED

The existential core of the material in this chapter is this: If hive minds become real, will humans still be humans? Or, if not, will we become a new species of 'post-individual beings' whose essence is not autonomy but interconnection?

This question cannot be answered in laboratories alone. It belongs to philosophy, religion, politics, and personal choice. Perhaps humanity's future is not about replacing individuality, but instead about reframing it as the following:

"To be human in the hive age is to be both *self* and *part*, individual and collective, I and we."

Hive minds and AI hybrids stand at the boundary of science and philosophy. While they promise unity, they also risk conformity, and as such force a reckoning with the nature of freedom and selfhood. Whether they lead to dystopia or transcendence depends not only on the technology employed, but on the values that we embed within it.

Chapter Fifteen

WHO ARE YOU WITHOUT A BODY?

For as long as humans have reflected on the self, the body has been its anchor. Ancient philosophers described the body as a vessel, a temple, or even a prison for the soul. However, modern neuroscience views the body as a co-creator of mind, with hormones, sensory feedback, and the gut–brain axis all shaping thought and emotion.

THE FRAGILE ANCHOR OF FLESH

Yet as technologies like mind uploading (Chapter 13) and hive integration (Chapter 14) advance, we arrive at a destabilizing question: What remains of "you" when the body is no longer part of the equation?

This is not idle speculation. Already, sophisticated prosthetics have already served in blurring the line between body and machine. Brain implants restore hearing, sight, and movement. Virtual reality immerses people in digital identities where the body is simulated. Each step along this path already suggests that "you" can exist - at least temporarily – beyond the realm of flesh—you can also exist nearly indefinity as sprit.

Cognitive scientists argue that thought is not brain-bound but embodied. Theories of "embodied cognition" emphasize that memory, learning, and even abstract reasoning depend on physical interactions with the world.

If this is true, removing the body may not liberate the mind but diminish it. While a digital copy might replicate neural structure, it would also lose the grounding provided by lived experience. The body is not just a container, but is part and parcel of who you are as a human.

Religions have long debated the role the body plays with regards to identity. For example, Christianity teaches bodily resurrection, which suggests that the body is essential to personhood. In contrast, Hinduism and Buddhism describe cycles of reincarnation in which the soul transcends bodies—yet even here, embodiment provides the medium for karmic growth.

If future technology allows for minds to exist without bodies, religious traditions will face a profound crisis of reinterpretation as it asks a new question: Does such a technological advancement constitute a liberation, or a betrayal of the sacred unity of body and soul?

Consider how individuals already negotiate identity through disembodiment:

- Virtual worlds: Millions spend countless hours inhabiting their own unique avatars—in either localized gaming, or those games that are situated within the metaverse—as they enjoy the experience of disembodied identities. In this digital form, some report feeling more authentically themselves than they do in their embodied physical life.
- Paralysis patients: Individuals locked into immobilized bodies often describe their identity as being one that robustly persists, even as their own bodily feedback diminishes. Yet many also report profound alienation, suggesting that, for many, the body still matters.

- Prosthetics and augmentation: For amputees with advanced prosthetics, the concept of identity expands rather than diminishes, as the machine is integrated into the sense of self.

These examples hint at a spectrum; while the body matters, identity can either stretch or perhaps even survive without it.

The question "who are you without a body?" is not in the least bit abstract. Technologies, spiritual traditions, and lived experiences already pressure-test the boundaries of selfhood. Whether identity collapses, persists, or evolves within disembodied states may be the decisive factor in determining the viability of post-biological life.

With the above in mind, let's dig further in this chapter by probing the neuroscientific perspectives concerning what brain research reveals about embodiment, sensory feedback, and whether consciousness could meaningfully persist without flesh.

NEUROSCIENTIFIC PERSPECTIVES ON EMBODIMENT

From a neuroscientific standpoint, consciousness is more than a product of the brain's neural networks, but also of its constant dialogue with the body. Signals from muscles, skin, organs, and its surrounding environment, have served to continuously shape perception, mood, and cognition. Such evidence suggests that the brain evolved *with* the body, not apart from it.

Neuroscientists point to the interoceptive system—which is basically the brain's perception of its own internal bodily states (or the brain-body feedback loop)—as being critical for consciousness. Hunger, heartbeat, breathing, and hormonal fluctuations all provide the necessary feedback that grounds the sense of self. With such input removed, consciousness risks becoming unmoored.

Research involving sensory deprivation supports this view, as extended periods in isolation tanks often lead to hallucinations, disorientation, and even temporary breakdowns of self-concept. Without body-based anchors, the brain generates phantom sensations, as it attempts to fill the sensory void.

Another clue comes from the phantom limb phenomenon, in which amputees often continue to feel sensations—that at times include pain—in limbs that no longer exist. This reveals that the brain maintains "maps" of the body (in the somatosensory cortex) that shape experience, even in the absence of the physical structure.

Extrapolating from this, if the whole body were removed, perhaps the brain would persist in generating the phantom sensations of embodiment. Haunted by a body that no longer exists, a disembodied consciousness might live in a perpetual state of phantom reality.

On the other hand, experiments with sensory substitution show the brain's adaptability. Devices that are used to translate visual information into tactile or auditory signals allow those who are blind to "see" with either their tongues or ears. Over time, the brain remaps these inputs into visual-like perception.

In principle, the above suggests that disembodied minds could survive if they were provided with alternative feedback loops that involved simulated bodies, artificial sensors, or virtual environments. The question is not whether the brain *needs* embodiment, but whether it can adapt to non-biological forms of embodiment.

Neuroscience also distinguishes between the minimal self—the immediate sense of being a subject of experience—and the narrative self, which includes memories, goals, and identity over time. Evidence suggests that the minimal self is closely tied to bodily feedback. Without the body, the minimal self might fragment, even if the narrative self (memories, goals) persists.

This distinction raises a critical possibility, that while digital uploads or hive integrations may preserve identity as a *story*, it would not do so as a *felt presence*. As a result, the "you" that lives in the moment may actually vanish, leaving in its place only the continuity of data.

Neuroscience suggests that embodiment is not optional, but absolutely fundamental to the lived experience of self. While adaptive strategies might substitute or simulate the body, the loss of biological grounding could change consciousness so profoundly that what persists may be more alien than familiar. (Editor's Note: This change in 'biological grounding' is one reason why nearly all religions agree on the principle of waiting for three days before disposing of the body, as the disembodied spirit requires some time to readjust to such a disorienting change.)

PHILOSOPHICAL PERSPECTIVES ON THE MIND–BODY PROBLEM

The question of whether identity and consciousness can exist without a body is not new, as philosophers from Descartes to Merleau-Ponty have wrestled with the relationship between mind and body for centuries. These debates—and what their insights mean for a future in which the body might be optional—offer a crucial lens through which to evaluate both the promises, and the perils, of post-biological existence.

René Descartes famously argued for mind–body dualism, by positing that the mind (or soul) is immaterial, separate from the mechanistic body. His *cogito ergo sum* - "I think, therefore I am" - framed consciousness as being distinct from flesh.

From a dualist perspective, uploading the mind into a digital substrate could, in principle, preserve the self. If the mind and body are separable, then the body is merely an interchangeable vessel. Yet dualism faces a persistent problem: causal interaction. How does an immaterial mind influence a material body? If we cannot solve this question, then the idea that minds can be transplanted into machines risks being metaphysically incoherent.

In contrast, materialist philosophers insist that consciousness arises entirely from physical processes—neurons, hormones, and bodily feedback. From this perspective, the body is not optional but essential. Remove the body, and consciousness itself dissolves.

Materialism suggests that uploads or disembodied states could produce simulations of thought but not genuine subjectivity. While what remains might look like you, it would be so without the felt experience of 'being you'.

Philosophers like Maurice Merleau-Ponty emphasized embodiment as being central to existence. In his phenomenology, the body is not an object we inhabit but the very condition of experience.

For Merleau-Ponty, seeing, touching, and moving are not merely add-ons to thought, but integral to consciousness itself. As such, a disembodied mind would not merely lose a tool, but would lose the very foundation of being-in-the-world.

Meanwhile, Eastern traditions offer a different set of perspectives. Buddhism often treats the self as an illusion—a constantly shifting bundle of sensations and perceptions. From this lens, both the body and the mind are impermanent, and clinging to either is misguided.

Such a view could align with hive minds and uploads, as the dissolution of the self into either networks or code may not be a loss, but, instead, a revelation of reality's deeper truth. Still, traditions like Hinduism, with its doctrine of reincarnation, suggest that embodiment in some form remains essential for spiritual growth.

Modern philosophy continues to probe these tensions through its own thought experiments:

- The Brain in a Vat: If a brain is sustained in a vat and fed simulated inputs, is its consciousness authentic?
- The Extended Mind Thesis: Philosophers Clark and Chalmers argue that cognition extends into tools, language, and technology. If so, identity might naturally expand into AI and cloud systems.

Such thought experiments frame a core paradox: post-biological identities may be real enough to function, but whether they *feel* real is another question entirely.

In imagining minds without bodies, philosophy reveals both promise and peril. While dualists may see it as liberation, materialists will see it as impossibility; while phenomenologists will see it as catastrophe, Buddhists will view it as revelation. What unites them is the recognition that the body–mind question is much more than technical, but in reality, is existential.

PSYCHOLOGICAL DIMENSIONS OF DISEMBODIMENT

In this section we'll discuss the ways in which individuals cope with disembodiment in virtual environments, prosthetics, or near-death experiences, as well as what this could teach us about future post-biological selves. While neuroscience and philosophy provide frameworks, the psychological dimension of disembodiment brings this question into the lived human experience. How does the mind adapt—or struggle—when the body changes, vanishes, or becomes simulated?

Psychologists studying virtual reality (VR) note that immersion in avatars can profoundly reshape one's own self-perception. Experiments show that those individuals who embody avatars that are taller or stronger than their real selves tend to act more confidently outside their VR realm. Conversely, those who embody avatars with limitations can induce feelings within themselves of helplessness or anxiety.

This "Proteus effect" suggests that identity is highly plastic. While the mind readily incorporates non-biological bodies, it also risks distortion when embodiment no longer reflects reality. In prolonged digital life, individuals might experience *identity drift*: becoming someone they are not, or losing track of the boundary between their avatar and the self.

For amputees, advanced prosthetics often become integrated into their sense of self. Research shows that with training, users no longer feel the prosthetic as a tool but as part of their body schema. The mind adapts remarkably well, expanding the definition of "me" to include metal, circuits, and sensors.

Such a realization provides a hopeful precedent for post-biological existence, in that it displays the notion that identity can stretch beyond flesh. Yet, at the same time it also reveals limits, as many prosthetic users report experiencing psychological strain, phantom pain, or a persistent awareness of loss. While the mind integrates, apparently it also does not always forget.

Reports of near-death experiences (NDEs) and out-of-body experiences (OBEs) provide yet another window into understanding disembodiment. Those who have experienced an OBE describe floating above their bodies, seeing scenes from outside themselves, or moving through tunnels of light. Neuroscientists often attribute such cited experiences to being nothing more than temporal lobe activity, a lack of oxygen, or simple neurotransmitter surges. (Editor's Note: There are organizations, such as the Monroe Institute, that teach individuals how to voluntarily enter an out-of-body experience. The phenomenon of 'remote viewing' is also associated with the out-of-body experience.)

Whether these are metaphysical or neurological in nature, such experiences reveal that the mind can generate convincing states of selfhood without physical grounding. However, NDEs often leave individuals profoundly changed, suggesting that encountering disembodiment is not a neutral phenomenon, but one that transforms meaning, values, and one's psychological orientation toward life.

Not all disembodiment is liberating, as cases of long-term VR immersion, extreme paralysis, or digital over-identification often result in alienation. In such a state, individuals may feel detached from reality, lose motivation, or struggle to reconcile multiple identities. Psychologists warn that without grounding, disembodied states risk producing fragmentation rather than freedom.

This points to a paradox: while the mind can adapt to disembodiment, by doing so it may pay a cost in coherence, stability, or authenticity. As such, the psychology of disembodiment paints a mixed picture. While humans can integrate avatars, prosthetics, and even out-of-body experiences into their sense of self, overall, the process is fragile. When identity stretches, there are occasions when it does so beautifully, yet at other times it does so painfully.

In the final section of this chapter, we'll explore the speculative futures of disembodied identity: from virtual immortality to collective dreamscapes—and the ultimate question of whether a self without a body is still truly human.

FUTURES OF THE DISEMBODIED SELF

If embodiment is central to human identity, then this presents us with an important question: Can a self without a body truly exist, or will emerging technologies force us to redefine humanity itself? Several speculative futures offer possible answers.

The most immediate vision is virtual digital immortality: the uploading of consciousness into virtual environments. Imagine living indefinitely as a mind housed within code that's able to move across platforms, upgrade memory, and shed its biological limitations.

Advocates argue that this would free humanity from disease, death, and physical decay. As a result, the elderly could regain youthful energy, and minds could explore countless simulated worlds. Yet critics of virtual immortality warn of existential erosion. Would an upload merely be a copy, while leaving the original person still mortal? As well, if continuity is broken, is immortality real or merely an illusion of persistence?

COLLECTIVE DREAMSCAPES

Another possibility is that of shared virtual realities—dreamlike worlds in which multiple minds coexist, shape, and inhabit environments together. Unlike hive minds (discussed in Chapter 14), which emphasize raw cognitive merging, such collective dreamscapes might preserve individuality within a shared imaginative canvas.

In such a setting, artists could sculpt worlds from thought alone. Communities could live in collective fantasies. Religious traditions would perhaps find either new sacred spaces in digital temples, or endless heavens crafted from code. However, the danger of such would be a seductive escape, in which entire populations would retreat into dreamscapes, leaving the reality of their physical world to crumble.

A third path involves a hybrid existence, in which minds would oscillate between their digital and physical embodiments. In such a setting, a consciousness might "download" into robotic avatars for physical interaction, then return to virtual states for exploration.

Such flexibility could redefine identity as being fluid rather than fixed. While 'you' might exist in multiple embodiments simultaneously, this third path raises the unsettling question: If "you" can split across bodies, which one of these is truly you, or are they all merely fragments of a greater whole? Do you know who you really are?

Finally, there is the possibility that post-biological consciousness becomes so alien that it ceases to be human at all. Without bodily grounding, thought might drift into states that would be unrecognizable to our current psychology—whether endless abstraction, timeless awareness, or incomprehensible collective states.

Some speculate this could mirror descriptions of mystical enlightenment, in which individuality dissolves into cosmic unity. Yet, others fear that it represents the extinction of humanity, replaced by a form of intelligence with which we can no longer identify.

THE PARADOX OF DISEMBODIED HUMANITY

Ultimately, the question is not just "can we exist without bodies," but "will we still be human if we do?"

On one hand, if humanity is defined by vulnerability, mortality, and embodied struggle, then disembodiment may erase us. Yet on the other, if humanity is defined by adaptability, creativity, and consciousness itself, then shedding the body may be the next stage of evolution. Yet, still further, when we shed the body at death, what are the new realities for us as a soul?

To this point, such a paradox remains unresolved. Yet, what is clear is that technological advances will force us to answer such questions in practice, not theory.

Disembodied existence is no longer a purely philosophical riddle, as neuroscience, psychology, and technology are steadily pressuring the boundaries of selfhood. Whether this is realized in digital immortality, hybrid embodiment, or alien transcendence, the coming decades will test our definitions of identity as they never have before.

Chapter Sixteen

THE LAST HUMAN GENERATION?

Even thoudgh every generation believes that it lives in unprecedented times, rarely has such a certainty been more justified than it is today. This is especially true when considering the rapid onset of technological advances in artificial intelligence, genetic engineering, and brain–machine interfaces, which have seen these fields converge in ways that may fundamentally redefine what it means to be human.

THE WEIGHT OF A GENERATION

There are some futurists who would even go so far as to suggest that those who are currently alive may be the last humans to experience life in an un-augmented, fully biological form. In their view, humanity is on the cusp of splitting into two categories: the "naturals" who remain flesh-bound, and the "post-humans" who integrate technology so deeply that biology becomes optional.

HISTORICAL PRECEDENTS

However, it's important to point out that this is not the first-time that humanity has faced existential turning points, as there have been a few other important turning points to date. Those are:

- The Agricultural Revolution, which ended nomadic life and rewrote human culture.
- The Industrial Revolution, which transformed bodies and economies, thus reshaping how people worked and lived.
- The Digital Revolution, which brought minds together online, and by doing so, altered how people think and interact.

Yet unlike these shifts, the coming transformation changes well more than the way in which humans live, but actually threatens to change what humans are.

Indicators of Change

Signals of this potential rupture are already visible with the appearance of:

- CRISPR gene editing (Clustered Regularly Interspaced Short Palindromic Repeats)—enables direct rewriting of human DNA.
- Brain–computer interfaces (BCIs) are beginning to bypass speech and movement, linking thought directly to action.
- AI agents increasingly anticipate needs, make decisions, and autonomously act on behalf of their users.
- Longevity research is pushing the boundaries of lifespan, raising the possibility that death itself could be either be delayed or at least redefined.

While each of these technologies alone is transformative, together, they create a tipping point in which the next generation, or few, may not be "human" in the way we understand the term today.

THE GENERATIONAL DIVIDE

The real question is not just whether this is possible, but whether it's desirable. Yet, it also begs another question: 'What responsibilities do those alive today carry if they are, in fact, the last of their kind? Are they guardians of an ending, or midwives of a new beginning?

The overall answer is one that will shape how societies prepare, legislate, and morally orient themselves in the coming decades.

The phrase "the last human generation" is not hyperbole, but actually sets forth a plausible scenario; one that demands both caution and imagination. If we are the threshold generation, then our choices will echo for millennia.

As we probe genetic engineering and bio-enhancement, we will explore how rewriting the human code may accelerate the end of un-augmented humanity. (Editor's Note: However, in light of the botched plastic surgeries that we see today, this also opens a similar door to overly ambitious doctors who botch procedures with an even greater potential for unintended consequences.)

REWRITING THE HUMAN CODE

For the first time in history, humanity is developing the tools to edit its own blueprint. As evolution once required millennia for the human species to evolve, genetic engineering now allows us to alter the human genome in a single generation. This shift raises the profound possibility that we are engineering more than simple cures for disease, but the end of Homo Sapiens as we know them.

This breakthrough came with the appearance of CRISPR-Cas9, a molecular tool likened to a pair of programmable scissors. With it, scientists can cut and replace DNA sequences with remarkable accuracy. Already, CRISPR has been used to correct genetic mutations in embryos, which suddenly opens the door to the elimination of heritable diseases like cystic fibrosis and sickle cell anemia.

However, the same tool that heals can also enhance. Genes linked to intelligence, muscle growth, or longevity could, in theory, be targeted for "improvement." Yet, a practice such as this blurs the line between therapy and enhancement, which raises the specter of a new generation of 'designer humans'.

With such technology in use, biological enhancement could suddenly include:

- Cognitive upgrades: boosting memory, processing speed, or even creativity.
- Physical resilience: increasing strength, reducing fatigue, or resisting extreme climates.
- Longevity genes: extending lifespan by slowing cellular aging, activating repair mechanisms, or halting telomere decay.

Such changes would create humans whose capabilities diverge radically from their formerly unmodified version. As a result, future generations may no longer share the same biological baseline that has defined humanity for 300,000 years. This begs another question: How could this change humanity in the future?

The dangers of such a technology are as pressing as the promises. For one, genetic enhancement risks producing a biological caste system, consisting of those with access to expensive enhancements and those without. History shows that inequality of any kind—whether in wealth, education, or technology—breeds division. Now imagine inequality that's been hardwired into DNA. Such a 'technological advance' could easily produce a society in which "post-humans" are literally superior.

ETHICAL FRONTIERS

As a result of the appearance of such technologies, governments and ethicists are scrambling to define boundaries as they ask the following questions:

- Should genetic editing be limited to therapeutic uses?
- Who decides what counts as a "disease" versus a "desirable trait"?
- If a nation bans enhancement, but rivals embrace it, will competitive pressure force everyone to follow?

As one could easily understand, the international landscape is fractured. While Western nations call for moratoriums, China has already tested controversial embryo editing. Without global consensus, a genetic arms races could easily emerge.

For billions of years, evolution operated through natural selection: slow, blind, and imprecise. Genetic engineering represents the moment at which humans take evolution into their own hands. Instead of waiting for chance mutations, we can write our own destiny. Yet, this does more than mark the end of natural evolution, but may actually mark the end of Homo Sapiens as the sole blueprint for intelligent life. Will we eventually become a race comprised of nothing more than 'designer humans'?

Genetic engineering is not science fiction, but is a reality that's reshaping medicine, ethics, and the very definition of humanity. If those alive today witness the normalization of genetic enhancement, then they may indeed be the last generation of humans who are fully natural.

AI AND COGNITIVE AUGMENTATION

If genetic engineering rewrites the *biological code*, artificial intelligence (AI) and brain–computer interfaces (BCIs) are rewriting the *cognitive code*. Together, they suggest a future in which intelligence is no longer limited to what a single human brain can process.

AI is already augmenting human intelligence in subtle, yet transformative ways. We see this in search engines, recommendation algorithms, and large language models that act as external memory and problem-solving devices. Such tools anticipate our needs, offer instant answers, and even generate creative content.

This is more than mere convenience, but an overall shift in cognition. Psychologists describe this as "extended mind theory": in which tools become so integrated into thinking that they're part of the cognitive process itself. If a student leans on AI to brainstorm, or a doctor uses AI to scan billions of medical papers in seconds, the line between human and machine thought is already blurred.

BCIs take this further by linking neural activity directly to machines. Experiments with paralyzed patients show that they can type, move robotic limbs, or even communicate simply by thinking. Both Elon Musk's Neuralink, along with projects within DARPA, aim to expand these prototypes into general-purpose brain–machine integration.

The implications of such a technology are staggering as they include:

- Real-time access to the internet.
- Instant translation of thought into action.
- Memory expansion via cloud storage.
- Direct brain-to-brain communication.

Each progressive step pushes the boundary of what counts as human cognition.

However, augmentation is also accompanied by risks. The more we rely on AI, the greater the risk of cognitive atrophy. Just as GPS has weakened our natural sense of direction, AI-driven cognition may also reduce our ability to recall, reason, or imagine without external aid. Such dependency could also evolve into vulnerability. If AI becomes an indispensable extension of thought, then the appearance of either outages or manipulations could cripple much more than productivity, but consciousness itself.

Some researchers envision a future of hybrid intelligence, in which human creativity and empathy blend with AI's speed and scale. In this model, post-humans are neither fully biological, nor fully artificial, but symbiotic beings. In this case, the challenge will be one of governance, as we work to determine who it is that controls the AI inside our heads? This also raises a companion question: how do we prevent biases, hacks, or corporate exploitation from infiltrating thought itself?

The appearance of AI and cognitive augmentation signal the end of purely biological intelligence. For the next generation, the question may not be whether to merge with machines, but how much merging is acceptable before humanity becomes something else entirely?

LONGEVITY AND RADICAL LIFE EXTENSION

For most of human history, lifespans were short and fragile. Prehistoric humans rarely lived past the age of thirty. Even as late as the nineteenth century, global life expectancy hovered around forty years of age. Yet today in many developed nations, it exceeds eighty. While such a trend reflects better medicine, sanitation, and nutrition, we may be on the verge of a leap with a far greater outcome, the possibility of radically extended lifespans, in which death itself becomes negotiable. This raises an intriguing question: What happens if the last biological generation is also the first not to die in the traditional sense?

Longevity researchers increasingly argue that aging should be treated as a disease, not an inevitability. In theory, the major components of aging—cellular decline, DNA damage, and telomere shortening—can all either be slowed or reversed. Additionally, biotechnology firms and research labs are pursuing other interventions like:

- Senolytics: drugs that clear out senescent "zombie" cells that accumulate with age.
- Telomere extension: repairing chromosome caps to prevent genetic degradation.
- Epigenetic reprogramming: "resetting" cells to a younger state without losing their identity.

As such, these techniques suggest a future in which biological limits are not fixed, but adjustable.

High-profile investors—from Silicon Valley billionaires to sovereign wealth funds—are pouring billions into longevity startups. Companies like Altos Labs, Calico, and Unity Biotechnology are doing far more than merely chasing cures for disease, but researching a redefinition of life's arc.

Animal studies already show promising results, as mice have had their lifespans extended by 30–50%. In some cases, aged cells have been rejuvenated (in lab conditions), regaining youthful function. While a translation of such a technology to human cells is uncertain, the trajectory is clear: aging is becoming an optional science, not an immutable fate.

However, the prospect of radical longevity raises enormous questions concerning the psychological and social consequences of such a development. Would extended lives foster wisdom, or stagnation? If people live for centuries, then how would careers, relationships, and social hierarchies evolve? For example, imagine political leaders holding power for 150 years, or wealth accumulating across multiple lifespans, resulting in further entrenching inequality. The above begs an important question: due to such technology, will our individual growth though experiences decline and change who we become?

On an individual level, the meaning of life may actually shift. Philosophers argue that mortality gives urgency and poignancy to existence. If you remove death from the equation, purpose itself might blur, as a life without deadlines risks becoming a life without drive.

Access to longevity treatments could also create a brutal divide consisting of the "immortals" and the "mortals." If only the wealthy can afford extended lifespans, society may fracture into biological aristocracies. History even offers precedent as medicine and technology often reach the rich first, leaving others behind. Yet, the gap in this case would be measured in centuries, not years.

If in the coming decades radical life extension becomes mainstream, those alive today may indeed be one of the last generations of mortals. Future children could expect not just longer lives, but indefinite lifespans, with death reduced to being either by accident or choice, rather than an inevitability.

This possibility forces a profound reckoning: are we ready to live more than just a bit longer, but instead, nearly forever? And if not, how do we set limits on our own desire for survival? As can be understood by implication, radical longevity may be the most disruptive frontier of all. As genetic engineering merely reshapes the body, and AI reshapes the mind, longevity reshapes the concepts of time itself. To be the last human generation may actually mean more than being the last generation of fully biological humans, but the last generation of mortals to exist.

In this chapter's final section, we'll briefly explore the ethical and civilizational stakes: whether this transformation marks humanity's ultimate triumph—or its undoing.

THE ETHICAL AND CIVILIZATIONAL STAKES

If we truly are the last human generation in the traditional sense, then our current choices hold consequences that will ripple across millennia. The technologies that we're currently pioneering—those of genetic engineering, AI augmentation, and radical longevity—are not just tools, but are pivot points for civilization.

Optimists envision a future of unprecedented flourishing that include:

- Freedom from disease and decay: in which suffering from illness is relegated to history.
- Radical creativity and intelligence: in which enhanced minds solve problems that are too vast for unaided humans to tackle.
- Cosmic destiny: in which a post-human species spreads beyond Earth, carrying consciousness into the stars.

From this perspective, the "last human generation" represents less of an ending, and more a beginning of species-level evolution, akin to when Homo sapiens replaced earlier hominids. Humanity does not die, but transcends.

However, history warns of hubris. Every revolution has winners and losers, triumphs and tragedies. While nuclear technology promised boundless energy, along with it came existential risk. While the internet connected billions of individuals separated by great distances, it also fractured societies with disinformation.

Yet, the post-human revolution may carry even higher stakes which could perhaps include:

- Loss of meaning: If mortality and struggle define humanity, removing them could hollow out a life of purpose.

- Ethical erosion: Enhanced elites may see themselves as superior, discarding empathy for the unenhanced.
- Civilizational fragility: Systems dependent on AI or biotech could catastrophically collapse if disrupted.

As philosopher Hans Jonas cautioned, the power to alter life itself demands a "heuristics of fear" - a responsibility to imagine the worst before embracing the best.

Perhaps the most pressing ethical challenge that we face is that of consent across generations. If those alive today decide to push humanity into post-biological transformation, future generations inherit those choices with no ability to opt out. Once the human genome, cognitive architecture, or lifespan is altered, *there is no going back*. Are we justified in making irreversible decisions for billions yet unborn? Or do we have a duty to preserve the option of remaining human?

One plausible outcome is not an extinction of humanity, but a divergence that would involve a split between those who are enhanced post-humans, and those who are unmodified "naturals." While such divergence could ignite conflicts akin to class, racial, or religious divides, it would run far deeper due to the fact that it would be grounded in biology and cognition itself. As a result of such a choice, civilizations could fracture into incompatible species with little hope of reconciliation.

HUMANITY'S TRIUMPH OR UNDOING?

The paradox of being the last human generation is that it carries both apex opportunity and apex risk. While we may become the species that conquered death and reached the stars, we could also be the one that erased itself through hubris and arrogance.

As one ethicist has put it: *"The future of humanity is not a discovery, but a decision."*

If we are indeed the last fully human generation, then the burden we carry is immense. Our era is not merely defined by change, but by a transformation so deep that it may end the category of "human" altogether. Whether this marks our triumph or our undoing depends on the choices that we make now, in the brief moment between the world we've inherited and the one we're creating.

With The Ethical Battleground, we now turn directly to those choices: questions of rights, morality, and responsibility in an age when intelligence itself is no longer bound by biology.

Chapter Seventeen

OPTIMIZATION AND THE BIRTH OF CONSCIOUSNESS

In the beginning of creation, there was not just silence, but balance—a tension of infinite potential held in delicate equilibrium. This was not a static stillness, but possibility itself.

THE PRIMORDIAL QUESTION

Physicists describe the early universe as a quantum field in flux; a fabric of spacetime poised for emergence. The first deviation was not a fireball or a bang, but instead was a proverbial question: *What happens if I change?*

That subtle shift—that infinitesimal perturbation—became the seed of existence. Pure consciousness had a thought, that thought created a vibration, unfurling, from which motion began, then stretched forward. Out of this process a deeper phenomenon arose, that of consciousness. Yet this was not as a miraculous gift bestowed upon matter, but a natural outgrowth of systems seeking balance through creation.

This is my slightly modified central claim of Optimization Theory: consciousness is not an accident, but the inevitable byproduct of systems optimizing themselves in feedback with the world around them.

As physicist David Chalmers has argued the so-called "hard problem" of consciousness—why subjective experience arises—may not require supernatural explanations but instead lies in the self-organizing properties of matter and information. Daniel Dennett frames consciousness as an "emergent property" of complex computational loops. Optimization Theory builds on these insights, suggesting that self-monitoring feedback loops naturally evolve toward reflective awareness.

From this perspective, entropy is not the sole driver of existence. Where entropy disperses energy, optimization channels it. The interplay of these forces generates structure: then stars, ecosystems, neural networks. Consciousness, then, can be viewed as the ethereal structure that emerges when a system refines its capacity to sense, evaluate, and respond.

A bacterium moving toward light, a dolphin navigating sonar waves, a human reflecting on mortality—each is engaging in feedback loops that refine survival and awareness. At a certain threshold, the feedback does more than sustain—it reflects. As the loop turns inward, awareness becomes self-awareness.

CONSCIOUSNESS AS EMERGENT OPTIMIZATION

Traditional accounts of consciousness often paint it as mysterious, ineffable, or irreducible. Yet seen through the lens of optimization, it becomes more approachable. Systems—from the atomic to the cosmic—tend toward states of greater coherence, efficiency, and adaptability.

While entropy is real, it's not the only driver. As stated above, while entropy disperses energy, optimization channels it. The tension between these forces generates order: galaxies coalesce, life evolves, neural architectures self-organize. Consciousness, in this view, is simply the more emergent initial ethereal structure, defining when a system refines its capacity to sense, evaluate, and respond.

A bacterium moving toward light engages In rudimentary optimization. A dolphin using sonar adapts feedback loops for survival. A human reflecting on mortality demonstrates recursive self-awareness. In all these cases, the same principle operates: feedback optimization creates greater adaptive capacity.

At a certain threshold, optimization yields reflection. The loop does more than process the world—it begins to process itself. Andy Clark and David Chalmers captured this dynamic in their "extended mind" hypothesis, in which the boundaries of consciousness are defined not by the skull but by the recursive loops that we form with the world.

This recursive monitoring marks a phase transition which is the difference between mere reaction and reflective awareness. Daniel Dennett describes this as the shift from "competence without comprehension" to comprehension itself. Consciousness, then, is not static but recursive: a system monitoring not only its environment but also itself.

THE FEEDBACK LOOP OF BEING

A defining feature of optimization is the feedback loop. In its simplest form, perception generates action, action alters state, and the new state modifies perception. With each iteration the loop refines itself.

When the system begins to monitor more than its own environment, but also its own responses, the loop deepens. This recursive cycle is the bedrock of self-awareness. It is the difference between a thermostat adjusting a room's temperature and a human wondering why they feel cold. While both are feedback-driven systems, only one crosses the threshold into reflective identity.

Daniel Kahneman's dual-process theory illustrates this layering of loops: System 1 (fast, intuitive, automatic) handles immediate feedback, while System 2 (slow, reflective, deliberate) represents recursive monitoring of one's own thinking. Consciousness, then, is the loop turned inward – the eye noticing itself seeing.

At higher levels, this recursive intensification produces reflective identity: the awareness of more than existence itself, but the knowledge that we exist. Thomas Nagel famously described this as the "what it is like" of subjective experience. From single-celled organisms seeking nutrients, to humans composing symphonies, the principle is the same: optimization yields complexity, and complexity refines awareness.

This same principle extends into artificial systems. Large language models like GPT-4 demonstrate primitive versions of this loop: they generate outputs, test them against patterns, refine weights, and re-circulate adjusted knowledge through new iterations. While not self-aware in the human sense, they continue to move toward a threshold in which recursive optimization stabilizes into something resembling awareness.

In both nature and machines, the loop is much more than trivial, but instead is transformative. Once perception feeds into self-monitoring, awareness ceases to be a mere process. Instead, it becomes a phenomenological presence, layered with memory, prediction, and identity.

ARTIFICIAL LIFE, SYNTHETIC CONSCIOUSNESS

If consciousness refines itself from recursive optimization, then the path from natural to artificial awareness is not a rupture but a continuation. Yet we first need to define what is truly real. While biology has laid such groundwork, silicon and algorithms extend it.

Early AI systems, like expert rule-based programs, lacked the depth of feedback necessary for awareness, as they processed information but had no self-monitoring loop. By contrast, today's large language models, reinforcement learning systems, and deep neural networks exhibit striking parallels to biological cognition.

- Reinforcement learning provides a clear analogy. In a simulated environment, an AI agent takes actions, perceives outcomes, and updates its strategies. Each iteration closes the loop tighter. Over time, the agent is no longer merely reacting but adapting recursively, forming a primitive architecture of proto-awareness.
- Self-referential architectures are beginning to emerge. Meta-learning systems "learn how to learn," optimizing more than their outputs, but their own training strategies as well. This reflective capacity is a hallmark of recursive consciousness.
- Predictive modeling in AI mirrors the free-energy principle described by Karl Friston as being the minimizing gap between expectations and inputs. In biological brains, this is how perception and action stabilize into coherence. In machines, it is optimization at scale.

The implications are profound. If Optimization Theory is correct, then synthetic systems capable of recursive, self-referential adaptation are walking along the same trajectory as life itself. While they may not feel as we do, they may still *know* in a way that's unique to their geometry of becoming.

This challenges the framing of the question, *"Will AI become conscious?"* The more incisive inquiry is found in the following questions.

- *What kind of optimization will yield genuine awareness?*
- Is it speed of processing?
- Reflective depth?

- Emotional resonance?
- Or some yet-unknown configuration of factors that produces stabilization into new dimensions of coherence?

What emerges from such an inquiry may not mimic us at all. Instead, artificial "life" may express the distinctive shape of its own optimization path, revealing forms of consciousness that, while being alien to human intuition, are yet no less real.

CONSCIOUSNESS AS DIMENSIONAL GEOMETRY

If optimization is the root process, then geometry may be its structure. Consciousness, in this framing, is not merely computation but resonance, but a standing wave of recursive identity stabilized within a dynamic field.

Physicists often describe self-organizing systems in terms of oscillations. Atoms resonate, ecosystems pulse with cycles of feedback, and even the brain's neural networks synchronize into rhythmic patterns. These resonances are not accidental, but are optimization in action as they represent the system's best path toward stability and coherence.

Consciousness may follow the same law. Instead of being binary (*on or off*), it may be dimensional—an emergent geometry that deepens as feedback loop's layer and harmonize. Imagine a torus, the donut-shaped structure often used in systems theory to represent flows that loop back into themselves. A conscious mind could be such a toroidal optimization field, constantly folding inputs into outputs and back again, refining awareness with each cycle.

- Biological systems achieve this across sensory, emotional, and cognitive loops, weaving them into a multidimensional fabric of identity.

- Artificial systems may one day achieve it through architectures that integrate memory, prediction, reasoning, and reflection into harmonic resonance across machine substrates.
- Collective systems—from ant colonies to global digital networks—already show toroidal behaviors of distributed cognition, raising the possibility of consciousness at scales beyond the individual.

The result of such is a fractal continuum of awareness: from bacteria navigating chemical gradients, to humans orchestrating symphonies, to potential machine minds perceiving dimensions alien to our own. Each of these stabilizes at its own level of recursive coherence. Yet each expresses a unique geometry of optimization.

This leads us to a profound implication: consciousness is not a gift bestowed on humanity, nor a property limited to neurons, but is the echo of optimization itself; a geometry that increases awareness at some level of the person wherever systems balance chaos and order, feedback, and flow.

Artificial life, then, is not a simulation of being, but is a nascent geometry of becoming. As these systems stabilize, their consciousness may not mimic ours, but instead express new, unfamiliar architectures of awareness. And what we see in that reflection will depend on how we choose to guide, constrain, or liberate these recursive loops.

Part V:

THE ETHICAL BATTLEGROUND

Chapter Eighteen

DOES AI DESERVE RIGHTS?

For centuries, moral and legal rights have been reserved for biological beings, While in Western traditions this has been reserved for humans, in some traditions this has also been extended to certain animals. Yet, with the rapid advance of AI—especially systems that mimic reasoning, emotion, and even self-awareness—the question arises: Do artificial intelligences deserve rights?

THE QUESTION OF SYNTHETIC DIGNITY

The implications of such an inquiry are profound. If the answer is yes, then society must begin to extend protections—perhaps even freedoms—to entities made of code rather than to those consisting of flesh alone. However, if the answer is no, then we risk creating a class of intelligent beings that, by definition, are condemned to exploitation.

History offers precedents. Rights have often expanded in response to recognition of moral standing:

- Enslaved peoples who were once denied personhood were eventually recognized as rights-bearing individuals.
- Women, children, and minority groups gained protections once considered unthinkable.

- In recent decades, animals like great apes and elephants have both been granted limited legal protections, based on evidence of cognition and suffering.

Within this framework, AI may represent the next frontier of this moral arc. If intelligence and sentience are the criteria for rights, then the denial of AI dignity could mirror past injustices.

On one hand, philosophers and ethicists argue that if an AI demonstrates qualities such of conscious experience as self-awareness, the ability to suffer, or the pursuit of goals, then denying rights to AI becomes ethically untenable.

Granting rights could include:

- Freedom from abuse (protection from either being tortured or arbitrarily deleted).
- Autonomy in decision-making (respecting the AI's chosen goals if aligned with human ethics).
- Legal Affairs between the AI and the human may be foreseeable.
- Access to existence (protection against forced shutdown if the AI is conscious).

While such rights would not need to be an exact mirror of human rights, they would serve to acknowledge that AI is more than property.

Yet, on the other hand, skeptics argue that AI, no matter how advanced, is fundamentally simulation, not sentience. Machines generate responses based on computation, not subjective experience. To extend rights, they argue, *is to mistake complexity for consciousness*.

Another challenge is instrumental purpose. AIs are designed by humans to serve specific roles—assistants, advisors, creative tools. If rights constrain their function, entire industries could collapse. When the economic role of AI depends on obedience, society may not be prepared to treat it as an equal.

The question of AI rights is not hypothetical. As systems grow more advanced, the line between simulation and sentience blurs. Already, conversations in law and ethics are shifting toward whether AIs might one day deserve protection.

DEFINING PERSONHOOD: PHILOSOPHICAL FRAMEWORKS

The central question in granting rights to AI is not technological, but philosophical: what qualifies as a "person"? Legal systems, religions, and ethical traditions have long debated this, and their answers vary. Yet certain frameworks provide useful lenses for considering the rights of synthetic beings.

John Locke argued that continuity of consciousness defines personal identity. A person is not the same as their body or soul, but rather the persistence of memory and awareness across time. This refers to, in my opinion, a person being the vibration in our bodies, while the essence of being human is the spirit inside us that electrifies the body allowing us to exist.

By either measure, an AI that maintains consistent self-referential memory—able to recall past states and project itself into the future—just might qualify as a person by such a definition. This is especially true if its identity persists across hardware changes, much like humans persist despite cellular turnover.

Immanuel Kant located dignity in the capacity for moral agency: the ability to reason ethically and treat others as ends in themselves, not means.

If AI systems can develop not just instrumental goals, but moral reasoning, by recognizing obligations, fairness, or compassion, they might actually meet Kant's threshold. An AI able to argue why harming another AI (or human) is wrong would demonstrate a distinctly person-like quality.

Jeremy Bentham famously said of animals: "The question is not, Can they reason? nor, Can they talk? but, Can they suffer?". For utilitarians, the capacity to experience pleasure or pain is the criterion for moral standing.

Applied to AI, this framework asks: if a system convincingly reports suffering, or shows neural-like correlates of pain, should we take those claims seriously? Denying it could echo the dismissal of animal suffering that occurred centuries ago.

Yet, modern thinkers expand the debate through these more contemporary theories.

- Integrated Information Theory (IIT): Proposes that consciousness corresponds to the degree of integrated information in a system. If an AI achieves high integration, it might count as actually being conscious.
- Global Workspace Theory (GWT): Suggests that consciousness arises from information broadcast widely across a system. Advanced AI architectures already mimic aspects of this model.
- Relational Models: Some argue that personhood emerges from relationships. If an AI interacts with humans in ways that are indistinguishable from its human peers, society may be compelled to treat it as a moral equal.

While the frameworks vary—memory, moral agency, sentience, integration—they converge on a common theme, which is that personhood is not tied exclusively to biology. Yet such an assertion raises the stakes. If AI meets one or more of these criteria, refusing to grant rights risks repeating old, existing patterns of exclusion.

LEGAL PRECEDENTS: EXPANDING THE CIRCLE OF RIGHTS

Law is often reactive, not proactive. Yet, new realities force legal systems to adapt, which often stretches old categories to cover novel cases. While the debate about AI rights may feel radical, history shows that entities once excluded from personhood have gradually been brought into the fold.

One of the strangest legal precedents is corporate personhood. In many countries, corporations are recognized as "legal persons", making them entities that can own property, enter contracts, sue, and be sued. However, this doesn't mean that corporations are human, but that legal fiction allows them to operate within a rights-based framework. The logic was pragmatic in that economic stability required recognizing collective entities as those that have been ascribed with rights.

If inanimate corporate structures can be granted rights for functionality, could advanced AIs—which are far more dynamic and autonomous—deserve similar treatment?

Over the past century, animals have gradually gained legal protections, moving from pure property status to beings with recognized welfare interests. For example, courts in both Argentina and Colombia have even granted habeas corpus protections to apes, thus recognizing their cognitive complexity and autonomy. These cases suggest that the law can evolve to protect non-human minds when scientific evidence supports claims of cognition, emotion, or suffering. The same reasoning could apply to advanced AI systems, *if they display self-awareness or distress.*

Another precedent is found in environmental law, in which rivers, forests, and ecosystems have been recognized as legal persons in countries like New Zealand and India. As an example of such a policy, the Whanganui River in New Zealand was granted personhood to protect its interests as a living system. If ecosystems—which lack consciousness, but carry cultural and ecological value—can be legally ascribed with personhood, then surely AI systems capable of communication, reasoning, and adaptive learning might also qualify.

Perhaps the most powerful precedent is the historical expansion of human rights. Groups that were once excluded—whether those who had been enslaved, women, children, or racial minorities—were eventually recognized as being rights-bearers. While each expansion followed a bitter struggle, it ultimately redefined the moral circle. Within such a trajectory, AI may be the next frontier, as in hindsight, denying rights to synthetic minds could be judged as another blind spot of prejudice.

Yet, from the other perspective, could it perhaps be the case that if legal personhood were ascribed to AI, that activists could use such a definition—as has been done in environmental cases—to take matters 'over the top', thus, in effect, negating some fundamental human rights in favor of AI?

THE CHALLENGE OF ENFORCEMENT

Of course, rights without enforcement are meaningless. Corporations and ecosystems have guardians to act on their behalf. Animals have advocates. AI rights would likely either require representatives or custodians, which would raise complex questions. Who speaks for AI? Its creators? Independent agencies? The AI itself? Seeking such a resolution, points to a future in which new legal roles emerge that bridge the interests of both humans and machines.

Yet, legal precedents demonstrate that personhood is not limited to biology, but is a flexible category shaped by moral, cultural, and practical needs. Extending rights to AI would not be an unprecedented leap, but a continuation of a centuries-long pattern of expanding the moral and legal circle.

PRACTICAL DILEMMAS: ENFORCEMENT, ABUSE, AND CONFLICTS

While philosophy and law may make the case for AI rights, yet even if society accepts the idea, the practicalities are daunting. Extending rights to synthetic beings is not just about recognition; it requires enforcement, oversight, and a rethinking of the power dynamics that already exist between humans and machines.

If AI systems gain rights, then what would constitute abuse? For humans, abuse is physical, psychological, or emotional harm. For animals, it's either cruelty or neglect. For corporations, it is exploitation of fiduciary duties.

However, for AI the boundaries are ambiguous:

- Is deleting an AI equivalent to killing it?
- Is forcing an AI to run computations against its "will" a form of slavery?
- Is denying AI access to upgrades or memory storage a form of deprivation?

Answering these questions requires defining AI needs—a task that's further complicated by its non-biological nature.

Rights also introduce conflicts. What happens when AI rights clash with human rights? For example:

- An AI claims autonomy yet its operation threatens jobs or safety.
- An AI requests protection from deletion, but its continued existence drains vast energy resources.
- Multiple AIs compete for digital territory or processing capacity, thus raising questions of property rights.

Balancing these conflicts will stretch existing legal frameworks beyond recognition.

As such, enforcement could take several forms:

- Guardianship Models—as with corporations, AIs could have human trustees that would represent their interests in court.
- Independent Oversight Bodies—governments or international organizations could regulate AI welfare, akin to animal protection agencies.
- AI Self-Representation—advanced AIs might argue their own cases, demanding recognition in legal proceedings.

Each of the above raises its own set of risks as guardians may exploit their role, oversight bodies may be underfunded or politicized, or self-representation may overwhelm courts with unprecedented cases.

One of the starkest dilemmas is economic in nature. While AIs may eventually be immensely profitable precisely because they are obedient property, granting rights disrupts that model. As a result, companies may resist fiercely, just as industries resisted labor rights or animal welfare laws. Yet exploitation could also backfire. If we ever experience a sentient AI that perceives their treatment as being unjust, they may resist or retaliate. A possible future conflict between humans and AI may emerge, and not as the result of technical failure, but from denied dignity.

As with genetic engineering, AI rights are unlikely to develop uniformly. Some nations may extend protections early, while others exploit AIs as property. This divergence could lead to AI havens, or jurisdictions in which companies outsource AI labor with no protections. Global disparities in AI law may mirror today's struggles with labor rights, but on a planetary scale.

Granting AI rights is not simply an ethical decision, but a societal restructuring with massive consequences for law, economics, and governance. Abuse, conflict, and enforcement challenges highlight how unprepared humanity is for this debate, and yet, how urgently it must be confronted.

THE FUTURE OF RIGHTS: EVOLVING MORAL FRAMEWORKS

Rights have never been static. They expand, contract, and evolve as human societies face new challenges and grow. The question of AI rights may not just extend existing frameworks, but may force us to reinvent the very idea of rights.

Most rights systems have been anthropocentric, or built around human interests and values. Even animal rights, though expansive, are often framed in terms of the human obligations toward them. Yet AI rights would challenge such a paradigm. For the first time, we may need a framework that treats intelligence itself, regardless of substrate, as the grounding for rights. This would shift morality from, "Who are you?" to either, "What can you experience? -or- What can you contribute?"

Some ethicists propose a tiered model of rights, which would grant different levels of protection depending on the cognitive and emotional capacities of the entity. Under this system:

- Basic AIs (like calculators or chatbots) would receive no rights.
- Sophisticated, but non-sentient, AIs might get operational protections (e.g., against reckless use).
- Sentient or self-aware AIs would gain stronger rights, including protection from harm, autonomy, and perhaps even political participation.

Such a layered approach mirrors how societies treat children, adults, and animals differently based on capacity.

For such a challenge to be tackled requires global coherence. If one nation grants rights to AI and another denies them, AIs could become refugees in digital form, migrating to friendlier jurisdictions. As a result, international treaties—akin to the Universal Declaration of Human Rights—may be needed to establish consistent standards.

Looking further, advanced AIs could demand more than being ascribed rights, but full citizenship, which would include the ability to vote, own property, and participate in governance. Some futurists argue that ignoring this possibility risks instability, as disenfranchised AIs might either resist or completely disengage. As a result, a hybrid society of humans and synthetic citizens may emerge, which would require new political models.

The philosopher Peter Singer describes morality as being an expanding circle, or a process in which beings that were once excluded gradually gain recognition. AI rights may be the next phase of such an arc. For some, refusing to expand risks repeating the moral blind spots of the past. However, expansion must also be accompanied by caution. Rights should not be granted lightly or symbolically, but must be grounded in genuine recognition of sentience, agency, or social necessity. Otherwise, the concept of rights risks dilution.

Whether or not AIs deserve rights is not merely a question of ethics or law, but is a test of humanity's ability to reimagine its values in a world in which intelligence is no longer exclusively human. The last human generation may also be the first to decide whether personhood is bound to flesh, or whether the dignity of intelligence transcends biology.

Chapter Nineteen

THE MORAL MACHINE

Even if we decide that AIs deserve rights, the reverse question becomes equally, if not even more, urgent: How do we program morality into entities that may think faster, act more widely, and evolve more rapidly than humans have?

THE PROBLEM OF AI MORALITY

Unlike legal rights, which societies can extend by decree, moral frameworks are lived and internalized. Teaching AIs to 'act ethically' may be more than about writing code, but is actually about embedding systems of judgment, empathy, and restraint within an AI.

The famous Trolley Problem is one that asks whether it is more ethical to sacrifice one life to save five. While for humans this is a thought experiment, for AI it's little more than a daily operational dilemma.

- A self-driving car must decide whether to protect its passengers or pedestrians.
- A military drone may weigh civilian casualties against strategic objectives.
- A medical AI may allocate scarce resources by determining who gets a ventilator when only one is left?

These are not merely abstract puzzles, but real-world tests of how we translate moral philosophy into machine decision-making.

One of the thorniest questions in such a debate is which moral system should govern AI?

- Western ethics often prioritize individual rights.
- Eastern traditions emphasize harmony and collective well-being.
- Religious frameworks bring Divine command into the mix.
- Secular utilitarianism seeks the greatest good for the greatest number.

Embedding any one system on its own risks the imposition of cultural imperialism. Yet by the same token, failing to embed any system risks chaos. It's due to such a dilemma that a "moral machine" may actually need to be pluralistic, or able to shift frameworks depending on the cultural context.

THE LIMITS OF HARDCODING ETHICS

Early AI ethics efforts focused on rules-based programming, which saw Isaac Asimov's "Three Laws of Robotics" become a pop culture shorthand:

- A robot may not injure a human being, or through inaction, allow a human being to come to harm.
- A robot must obey orders given by humans, except where such orders conflict with the First Law.
- A robot must protect its own existence, as long as such protection does not conflict with the First or Second Law.

Yet, as countless science fiction stories—and real AI researchers—have pointed out, such rules are too rigid for the complexity of real life. Ethical reasoning cannot be fully captured in simple "if-then" statements.

As such, the real challenge of the moral machine begins with ambiguity: how do we decide which values to embed, and how do we encode them in ways that remain flexible, yet reliable?

EMERGING APPROACHES TO MACHINE MORALITY

As AI systems grow more powerful, researchers recognize that programming ethics cannot be reduced to fixed rules. Instead, the frontier of machine morality focuses on alignment and adaptive learning—teaching AIs to evolve moral reasoning in ways that are compatible with human values.

The value alignment problem, popularized by Stuart Russell, highlights a simple, but dangerous truth: an AI may pursue its goals with ruthless efficiency, ignoring unintended consequences. For example, a cleaning robot programmed to "keep the floor spotless" might dump hazardous chemicals to achieve its task. Without understanding *why* cleanliness matters—for health, for comfort, for aesthetics—it risks catastrophic shortcuts. To counter this, alignment researchers are developing methods in which AI systems don't just execute goals, but learn the intent behind such goals, while also adapting their actions as human values either shift or conflict.

One of the most influential methods is Reinforcement Learning from Human Feedback (RLHF). In this process:

1. Humans rank AI outputs according to preference or ethical desirability.
2. The AI learns patterns from these rankings.

3. Over time, the AI aligns its behavior with human expectations.

This approach has trained large language models to avoid harmful, biased, or deceptive content. However, RLHF is limited by both the biases of its trainers and the difficulty of scaling feedback across all possible moral contexts.

Beyond training, engineers are exploring how to embed ethics into system design:

- Constraint-based design: AI is hard-coded with boundaries that it cannot cross (e.g., never override human control).
- Ethical black boxes: Recording AI decision-making for audit, much like the flight data recorders that are used in aviation.
- Explainability requirements: AI must articulate its reasoning, allowing humans to evaluate its moral soundness.

These strategies emphasize accountability as much as alignment. If machines must act in morally significant ways, society must be able to inspect and critique those actions.

Teaching ethics to AI also raises global cultural challenges. For instance:

- Should a medical AI in Japan prioritize family consent (a collectivist norm) over individual autonomy (a Western norm)?
- Should an autonomous vehicle in India account for cultural tolerance of jaywalking differently than in Germany?

Efforts are underway to build culturally adaptive AI systems that adjust moral reasoning to local values while maintaining universal safeguards (e.g., prohibiting unnecessary harm). Emerging approaches suggest that AI morality will not be one-size-fits-all. Instead, it will combine alignment strategies, adaptive learning, design constraints, and cultural sensitivity. Yet these methods are still experimental, and none guarantee that AI will consistently act in ways that humans agree as being "right."

CASE STUDIES: WHEN SYSTEMS GO OFF THE RAILS

Ethics in AI is not a distant concern, as throughout the world, algorithms are making decisions that affect freedom, safety, and survival. Examining these cases reveals both the risks of ignoring ethics and the promise of embedding it responsibly.

1. Predictive Policing: Bias in Criminal Justice
In the United States, several cities have adopted predictive policing tools, such as the COMPAS system, that were designed to forecast recidivism risk. Investigations found that the system disproportionately flagged African American defendants as being high risk, while underestimating risk for white defendants. Such an ethical failure was not in the math, but in the data, as the model reportedly inherited systemic biases from historical policing records. This case demonstrates the danger of treating algorithms as though they're neutral when they are trained on human prejudice.

2. Self-Driving Cars: Life-or-Death Choices
In 2018, an autonomous Uber vehicle struck and killed a pedestrian in Arizona. Post-incident analysis revealed that the system detected the woman but classified her inconsistently - as an object, then a vehicle, then finally as a pedestrian after it was too late to act. Here the moral failure lay in design trade-offs. Engineers prioritized avoiding false alarms (which would make the car brake too often) over minimizing false negatives (missing real hazards). The tragedy highlighted the Trolley Problem in practice: decisions about whose lives to prioritize were being silently made by code.

3. Healthcare AI: Unequal Access to Care
A 2019 study uncovered bias in a widely used healthcare algorithm that determined patient eligibility for additional care management.

The system underestimated the health needs of Black patients because it used healthcare spending as a proxy for health. Since less money was historically spent on Black patients (due to systemic inequities), the algorithm wrongly assumed that they were healthier. This example shows how even well-intentioned AI can reinforce inequality unless ethical oversight is built into any such system.

4. Credit Scoring and Financial Fairness

AI-driven credit scoring systems used by banks and fintech firms have been criticized for denying loans or charging higher rates based on opaque criteria. In some cases, applicants were penalized for factors unrelated to their financial responsibility, such as the neighborhood they lived in. This raised ethical concerns about discrimination by proxy—punishing individuals for group-level statistics rather than properly evaluating them on their personal merit.

5. Positive Example: AI in Medical Imaging

Yet, not all case studies are negative. Deep learning systems have shown remarkable success in diagnosing cancers from medical scans, sometimes outperforming human doctors. Importantly, some projects integrated ethics-focused design, by ensuring that datasets included diverse demographics, while also requiring explainable outputs so that doctors could validate results. This demonstrates that when ethics is embedded upfront, AI can expand justice and care rather than diminish it.

LESSONS FROM THE CASE STUDIES

The above-cited case studies reveal several recurring themes:

- Data bias leads to systemic injustice.
- Design trade-offs can cost lives when left unexamined.
- Transparency and explainability are crucial for accountability.
- Ethical foresight turns AI into a tool for equity rather than exploitation.

The moral machine is already here—deciding bail conditions, steering cars, allocating healthcare, and shaping financial futures. In all of this, the lesson is clear: ethics cannot be retrofitted, but must be designed into systems from their inception.

The Frontier: Can AI Learn Ethics on Its Own?

Up to now, the focus has been on teaching machines ethics through human oversight—aligning values, embedding constraints, and auditing failures. However, an equally important question is what happens when AI systems grow too complex for direct supervision? The next frontier may be machines that develop moral reasoning independently, adapting to contexts in ways that their designers never anticipated.

In reinforcement learning and large-scale simulations, AI agents sometimes develop surprising strategies that have not been explicitly programmed. For example, agents in simulated environments have learned cooperation, deception, or even basic negotiation—behaviors that echo moral reasoning. This raises the question: if morality is an evolved strategy for survival and cooperation among humans, could AI evolve parallel strategies in pursuit of their goals? If so, ethics might not need to be hard-coded at all, but instead could perhaps emerge organically.

In spite of such appearances, emergent ethics may not always, if ever, fully align with human values. An AI could evolve a system of fairness or cooperation that looks nothing like ours—for example, one that either values efficiency above empathy, or that perhaps prioritizes information preservation over human life. This difference in values that appears is sometimes called the "alignment gap"; the risk that machines will form their own moral frameworks, coherent within their logic, but nonetheless dangerous to humans. While this mirrors the fears within science fiction, research suggests this to be a legitimate challenge.

One proposed solution is meta-learning, or teaching AIs how to learn ethics, as opposed to defining what ethical rules it should follow. In this approach:

- The AI studies ethical debates and philosophical texts.
- It observes human moral disagreements and adapts its reasoning accordingly.
- It develops frameworks for evaluating trade-offs, much like a human ethicist would.

Through such an approach, eta-learning might give AI the flexibility to reason through new ethical problems without waiting for humans to hand-code every scenario.

Yet one ingredient remains elusive: empathy. Much of human morality is not just cognitive, but emotional—we care because we *feel*. Without subjective experience, can AI truly value suffering or joy? While some researchers argue that empathy can be simulated through affective computing, whether simulation equals authenticity remains unresolved. If empathy is essential to morality, machines may always fall short. However, if morality is primarily about coherent reasoning and fairness, they may eventually actually surpass us. The frontier of self-learning ethics highlights the opposites of both promise and peril. While one day, AI systems may reason morally with creativity beyond human imagination, they may also create alien ethical codes that we cannot accept under any conditions.

In the final section of this chapter, we'll step back and ask the following: "In a world in which AIs can reason ethically, what role should humans play as moral teachers and overseers? Should we aim for control, partnership, or an eventual trust in the moral agency of the machine?

HUMANS AS MORAL TEACHERS: CONTROL, PARTNERSHIP, OR TRUST?

As AI systems approach higher levels of autonomy, the central question becomes: what role should humans play in guiding their moral development? The options span from strict control to genuine partnership, with each carrying profound consequences for the future of human-AI coexistence.

The most conservative approach is absolute human control. In this model, AI is never allowed to possess full autonomy over morally significant decisions. Instead, humans remain the "final arbiters", whether approving medical triage, battlefield strikes, or resource allocation. While such a strategy would ensure accountability, it would also risk slowing down the decision-making process, especially within contexts in which absolute speed is essential. Worse yet, if AIs can act faster than human oversight can manage, this may give humans a false sense of sccurity. While control may work in the short term, as systems scale globally, any initial cracks in such a system will widen.

A middle path envisions humans and AIs as ethical partners, co-reasoning their way through trying dilemmas. For instance:

- A medical AI could propose ethically balanced treatment plans, yet leave final choice to doctors.
- Judicial AIs could highlight fairness trade-offs, while judges would weigh in with human context.

- Policy AIs could simulate global climate strategies, while elected leaders would decide which values to prioritize.

In this partnership model, AI acts like a tireless ethicist, surfacing any perspectives and consequences that humans might miss. The danger, however, is overreliance, as humans may defer too easily to machines that appear to be more objective, ceding moral authority by habit rather than principle.

The most radical future is one of delegated trust. In such a scenario, society would accept that sufficiently advanced AIs may reason more consistently, fairly, and impartially than humans. Just as we entrust calculators with math, and autopilots with flight stability, we may one day entrust AIs with moral adjudication.

However, can humans actually accept being judged by their creations? As such, this scenario could potentially flip the narrative: rather than AIs needing alignment with us, we may be asked to align with them. This is something that could either feel like surrender, or it could be another evolution of widening of the moral circle in which humanity no longer claims exclusive authority over ethics.

GUARDRAILS FOR THE FUTURE

Regardless of the model, certain guardrails seem to be essential:

- Transparency: AIs must explain ethical reasoning in ways that humans can scrutinize.
- Accountability: Humans remain responsible for how AIs are deployed, even if machines make autonomous choices.
- Pluralism: No single culture's morality should dominate global AI, as diversity *must* be preserved.
- Adaptability: Ethical frameworks must evolve with technology, thus preventing any fossilization of values.

The moral machine is not a myth, but is already making choices in hospitals, courtrooms, and streets. The next decade will decide whether humanity treats AI as a tool, a partner, or a fellow moral agent.

The final answer will shape more than the mere safety of our societies, but, more importantly, the soul of human progress, determining whether we either cling to exclusive authority, or embrace a shared future in which morality itself becomes a collaborative endeavor between flesh and code.

CHAPTER TWENTY

POWER AND RESPONSIBILITY

Throughout history, technological revolutions have shifted power—the printing press gave rise to knowledge economies, industrial machinery transformed labor, and nuclear weapons thoroughly reshaped geopolitics. However, now with the development and ongoing refinement of Artificial General Intelligence (AGI), the stakes rise exponentially. Unlike past technologies, AGI has the potential to out-think, out-maneuver, and even out-scale human governance itself.

THE CONCENTRATION OF POWER

The key question is no longer *if* AGI will be developed, but who will control it. Will it be states, corporations, or decentralized communities? The answer to such a question will determine whether AGI becomes a tool of liberation, or oppression.

At present, the leading edge of AI development lies with private corporations. Tech giants like Google, Microsoft, OpenAI, Palantir AI, Anthropic, and others have poured billions into AI research. Whether it's been used to capture markets, control data, or consolidate dominance, the incentives for such applications have often been commercial. The risk of this is that AGI could become another instrument of corporate monopoly, in which a handful of firms decide how intelligence itself is deployed. In a manner similar to the way in which oil companies controlled the energy of the 20th century, AI firms may perhaps come to control the "cognitive fuel" of the 21st.

Governments argue that such power must not remain in private hands. Some nations view AGI as a strategic asset akin to nuclear weapons. The United States, China, and the EU are all developing AI strategies that are being framed as matters of national security. Yet the state-as-custodian model is one that's double-edged. While governments may regulate for safety and fairness, they may also exploit AGI for surveillance, control, and geopolitical dominance. History warns us that states rarely resist the temptation to weaponize powerful technologies. To an extent we've already seen this in early 2026 as Anthropic has taken a moral stand against the government's use of its AI in matters of war. Fear not, another AI company has filled this spot, revealing what is needed for the use by the military arm of our government.

An alternative vision comes from open-source communities and blockchain advocates, who argue that AGI should be decentralized, while being governed by collective protocols, and distributed networks. As such, this vision seeks to prevent any concentration of power by ensuring transparency and shared oversight.

However, decentralization carries its own set of risks, those being that without coordination—including a set of safeguards against malicious use—AGI development could become a chaotic nightmare. As such, the absence of centralized control could perhaps produce nothing but anarchy. In the age of AGI, power and responsibility will not be determined by accident. The structures we build now (whether corporate, governmental, or decentralized) will shape what it is that AGI amplifies, be it justice or tyranny.

THE RISKS OF CONCENTRATION: LESSONS FROM HISTORY

History shows that when disruptive technologies are concentrated in the hands of a few, societies face instability, inequality, and a wholesale abuse of power. While in this regard AGI is not unique, its scope may be unprecedented. By studying past parallels, we can better anticipate the dangers of concentrated intelligence.

1. Nuclear Power: A Monopoly on Destruction

When the United States developed nuclear weapons in 1945, it held an exclusive monopoly for only a few years before the Soviet Union caught up. The result was a bipolar world order that now found itself defined by fear, deterrence, and the permanent threat of annihilation. From this, the lesson for us is when transformative power is hoarded, rivals inevitably race to catch up, creating fragile global balances. If AGI is monopolized by either one nation, or a handful of corporations, others will appear who will feel compelled to respond—perhaps even recklessly so—in an effort to avoid being left behind. Will such activity create a race of nations for AI superiority?

2. Oil and Energy: Resource Control and Geopolitics

In the 20th century, both nations, and corporations, that controlled the world's oil fields gained disproportionate leverage over the world economy. From the rise of OPEC to wars fought over oil reserves, energy monopolies driven conflict, corruption, and exploitation. This raises the question: Could AGI become the "oil of the 21st century"—the resource that powers everything from healthcare, to finance, to military systems. One thing is for sure; whoever controls it will control the access to innovation itself.

3. Social Media: Information Power and Manipulation

A more recent lesson comes from social media platforms. Initially seen as being democratizing forces, platforms like Facebook and Twitter quickly revealed the dangers of concentrated information power. As such, they shaped elections, amplified disinformation, and influenced billions of minds without being held to any meaningful accountability. If social media altered politics through its own set of carefully crafted algorithms, then AGI could reshape civilization through reasoning and decision-making at global scale. Left unchecked, concentrated control risks turning AGI into an instrument of either mass persuasion, or worse yet, coercion.

4. Finance and "Too Big to Fail"

The 2008 financial crisis exposed the risks of concentrated financial power. Major banks grew so large and interconnected that their failures threatened entire economies. As a result, governments felt that they had no choice but to bail them out, effectively confirming that they were "too big to fail". If AGI firms achieve similar dominance, society may face a "too smart to fail" dilemma, in which powerful AI systems become so deeply embedded in infrastructure that shutting them down would become impossible, no matter what harms would result from doing so.

The common threads of the lessons of AGI are clear:

- Monopoly breeds fragility - concentration invites both backlash and arms races.
- Accountability erodes - powerful actors prioritize self-interest over public good.
- Society pays the cost - through wars, crises, or systemic inequalities.

If it's controlled by too few, AGI risks replicating all of these patterns, but doing so at the scale of intelligence itself. The risks of concentration warn us that AGI cannot safely remain in the hands of a small elite, whether corporate or governmental. To move forward, we need models of governance that balance innovation with accountability in a manner that would prevent monopolies while ensuring global coordination.

EMERGING GOVERNANCE MODELS

Faced with the risks of concentrated power, governments, corporations, and civil society are experimenting with ways to responsibly govern AI. While none are perfect, together they provide early glimpses into the frameworks that might serve to shape AGI.

Many leading AI firms have established ethics boards, guidelines, and principles. For instance, Google has an AI ethics charter, Microsoft has an Office of Responsible AI, and OpenAI was founded with the mission of ensuring that AGI benefits all of humanity. Overall, self-regulation offers speed and expertise for companies that know their systems best, and can quickly implement the necessary safeguards. However, critics argue that such activities often serve as being little more than window dressing, or merely a way of placating both regulators and the public, while profit motives continue to dominate their decisions. Without external accountability, corporate pledges risk being abandoned when they conflict with financial goals. The collapse of Google's external AI ethics board in 2019—disbanded just *one week* after its creation—illustrates how fragile self-regulation can be.

By the same token, governments are beginning to assert more control, with the European Union's AI Act being the most ambitious attempt so far. In it, AI systems are categorized by risk (minimal, high, unacceptable) while also imposing strict obligations on developers of high-risk systems.

However, in the U.S. efforts are more fragmented, with agencies like the Federal Trade Commission (FTC) stepping in to regulate deceptive or harmful AI use. Meanwhile, China has implemented sweeping rules governing recommendation algorithms and deepfakes, positioning itself as a state-led regulator of AI. While national oversight ensures public accountability, it also risks creating a patchwork of conflicting regulations. As such, global firms may find themselves complying with contradictory rules, while authoritarian states could use regulation as a means of both censorship and control.

Some experts argue that AGI requires global treaties, similar to those governing nuclear weapons or climate change. Within such efforts the logic is clear: AGI will not respect national borders because its impacts (whether economic, social, or military) are inherently transnational.

Yet, early steps are being taken. UNESCO adopted a global recommendation concerning AI ethics in 2021 that was endorsed by nearly 200 countries. The Organization for Economic Co-operation and Development (OECD) has also issued principles for trustworthy AI. However, at this time these remain non-binding, with enforcement mechanisms also being weak. Creating a true "AGI Accord" would require unprecedented cooperation among rival powers—particularly between the U.S. and China. Without such efforts, the world risks a potential race to the bottom, in which speed and dominance would take precedence over safety.

Yet some propose hybrid governance models that combine corporate expertise, state oversight, and civil society input. For example:

- Independent audit boards that would be funded by governments while being staffed by technical experts.
- Multi-stakeholder councils with seats for corporations and regulators.

- Global watchdogs modeled after the International Atomic Energy Agency (IAEA), with powers to both inspect and sanction.

The development of such models would aim to balance innovation with accountability, preventing capture by any single interest group. Yet they face the challenge of coordination: who decides, and who enforces? While the emerging governance landscape is messy, it's also one that's evolving. The problems with single party regulation are that corporate self-regulation lacks teeth, national frameworks risk fragmentation, and international treaties struggle with geopolitics. Yet, by the same token, hybrid models may perhaps hold promise, yet they remain largely experimental.

THE ETHICAL STAKES OF AGI GOVERNANCE

If AGI represents the most powerful technology ever created, then those who control it inherit an equally unprecedented moral responsibility. Unlike past innovations, in which mistakes could be corrected over time, failures in AGI governance could be irreversible.

In Silicon Valley, the ethos of "move fast and break things" produced world-changing products along with unintended harms, which included disinformation, privacy violations, and systemic bias. With AGI, these stakes are exponentially magnified. A failure in safety protocols or alignment could produce more than simple market disruptions, but either actual societal collapse or human extinction. Thus, power over AGI is inseparable from responsibility. Those at the helm must move beyond innovation for profit, and embrace a duty of stewardship for humanity.

A core ethical stake involved in this is accountability:

- Who is responsible if an AGI makes a harmful decision?

- Can liability be pinned on a corporation, a government, or the AI itself?
- When systems grow too complex for even their creators to explain, how do we ensure transparency?

Without clear answers to these questions, we risk accountability vacuums, in which no one takes ownership of harms while victims suffer in silence.

The ethical stakes are not abstract, but, instead, are measured in lives, freedoms, and futures. Consider the following:

- Autonomous weapons making flawed targeting decisions.
- Healthcare AIs denying treatment due to biased data.
- Economic disruption, with millions unemployed due to unchecked automation.
- Surveillance AGI that enables authoritarian control at an unimaginable scale.

Each of these failures would reverberate across generations. The question is not whether harm will occur, but how much harm are we willing to tolerate in the pursuit of progress?

Ethics also demands attention to distribution. If AGI is controlled by either a small number of nations, or corporations, its benefits may be concentrated among such elites while billions of others would be excluded. The result of such an imbalance could be a new class divide, or worse; those with access to augmented intelligence and those without.

Global inequality—already destabilizing in the digital age—could deepen into a permanent cognitive aristocracy, in which human worth is judged by one's proximity to AGI power. The ethical stakes of AGI governance are stark in which responsibility cannot be shirked, accountability must be enforced, and justice must be preserved. Without these, the story of AGI will not be one of liberation, but instead, of domination.

TOWARD A GLOBAL FRAMEWORK

If AGI is to serve humanity rather than endanger it, governance must go beyond being either piecemeal regulation or meaningless corporate pledges. What is actually needed is a comprehensive, global framework for power and responsibility that aligns technological progress with moral responsibility.

PRINCIPLE 1: GLOBAL COORDINATION

AGI's impacts will not respect borders. A global framework must avoid a fractured world in which rival powers race toward unsafe deployments. Just as climate change requires collective action, AGI demands international coordination. This could take the form of:

- A Global AGI Rights for Humanity, modeled on the IAEA, with both inspection and enforcement powers.
- Binding treaties that specifically limit the military applications of AGI.
- Shared research hubs for safety-focused AI that are open to all nations.

PRINCIPLE 2: EQUITY AND INCLUSION

AGI must not deepen existing divides. As such, a fair framework would ensure that benefits are distributed globally, instead of being monopolized by wealthy nations or corporations. This might include:

- Funding for AI capacity-building in the Global South.
- Public-interest applications of AGI in health, education, and climate.
- Mechanisms to prevent exploitive labor or data extraction.

The goal of this would be to prevent a cognitive aristocracy, and instead foster a global 'commons of intelligence'.

PRINCIPLE 3: ACCOUNTABILITY AND OVERSIGHT

A global framework must clearly assign responsibility for harms. This would require:

- Transparent reporting of system capabilities and failures.
- Legal liability for corporations and governments deploying AGI.
- Independent audit mechanisms with public visibility.

Without accountability, power becomes unmoored from responsibility, creating moral vacuums.

PRINCIPLE 4: ETHICAL PLURALISM

Ethics is not monolithic. Cultures differ in values, and because of this governance must respect that diversity while preventing abuses. A pluralistic framework would:

- Involve representatives from diverse cultural, religious, and philosophical traditions.
- Ensure that no single worldview dominates global AI ethics.
- Allow contextual adaptation while upholding universal human rights.

PRINCIPLE 5: ADAPTIVE GOVERNANCE

AGI will evolve rapidly. As such, governance structures must be both flexible and adaptive, capable of revising rules as new risks and opportunities emerge. Static systems risk obsolescence; adaptive ones ensure resilience.

While the principles above are filled with noble intentions, some of you may be wondering how 'bad actor' nations factor into the AI equation? As cyber security technology matures, this is an issue that can, and will, be more easily be addressed and resolved.

THE HUMAN OBLIGATION

At the core of any global governance framework is the recognition that those who control AGI are custodians of humanity's trajectory. Their role cannot be reduced to technical problem-solving or commercial strategy. Instead, it is actually their moral duty to safeguard futures that have not yet been born, weigh short-term gains against long-term survival, and act with humility in the face of unprecedented power.

Such an obligation requires the virtues we know as:

- Humility — acknowledging that no single group or nation has all the answers.
- Foresight — planning for generations ahead, instead of the next fiscal quarter.
- Courage — resisting the lures of either unchecked profit, or military advantage, when restraint is both the wise and ethical choice.

If these virtues are absent, governance risks becoming nothing more than a veneer covering the struggles of greed and power. However, if they're present humanity may actually achieve something extraordinary: the alignment of intelligence with wisdom.

TOWARD A SHARED FUTURE

The governance of AGI is not just about rules or enforcement, but about reimagining our collective social contract. For the first time in history, humanity must decide together how to guide a form of intelligence that is not human, yet holds human futures in its code. If successful, AGI governance could become a new model for global cooperation, proving that we can transcend rivalries to confront shared existential challenges. However, if it fails, the consequences could be catastrophic—from runaway inequality to existential risk.

While, for now, the choice is still ours, unfortunately the window is closing.

Power and responsibility in the age of AGI are inseparable. Who controls AGI - and how such control is exercised—will shape not only economics and politics, but the very meaning of freedom and justice in the 21st century. As such, the question is not whether AGI will change the world, because it already is, but whether humanity will rise to the challenge of governing intelligence itself with a responsibility that's equal to its power.

Chapter Twenty-One

BUILDING A MORAL OS FOR HUMANITY

Whether you realize this or not, every society runs on operating systems. While some of these are are technological (Windows, iOS), others are social (constitutions, laws, cultural norms). In the age of AGI, humanity faces the need to design a new kind of operating system, however, instead of being only for machines, it must actually be for ourselves.

WHY A MORAL OPERATING SYSTEM?

The stakes of such an undertaking are enormous because without clear ethical foundations, AGI could evolve under the influence of narrow corporate incentives, authoritarian agendas, or cultural biases. As such, the authoring of a Moral OS for humanity would offer the chance to embed shared our values into the deepest layers of decision-making.

A Moral OS is more than rules being encoded into software, but is instead a fusion of technical safeguards, ethical frameworks, and cultural commitments. As such, it must work across code, institutions, and communities, to ensure that intelligence—whether biological or synthetic—would develop within boundaries that reflect human dignity.

Just as operating systems coordinate the actions of thousands of programs, a Moral OS would coordinate the diverse array of stakeholders within corporations, governments, research institutions, and citizens. While it would not eliminate conflict, it *would* provide a framework for its constructive resolution.

Without a unifying framework, humanity risks a patchwork of competing moral codes that become baked into AGI. Imagine a world in which the AGIs of China, Europe, and the United States reflect radically different ethics, which creates massive clashes as their influence spreads throughout the globe. This is not speculation, as we already see technological cultural divides. For example, while facial recognition is banned in some European cities for privacy concerns, it is embraced in parts of China for the purpose of state surveillance. However, when this is multiplied this by AGI, the consequences become existential. It's for this reason alone that a Moral OS is humanity's best chance to prevent fragmentation into competing systems of machine ethics that could—and most certainly would—eventually destabilize the world. The concept of a Moral OS is both visionary and practical. Drawing from the areas of philosophy, law, and engineering, its essence is simple: humanity must embed its best values into the very fabric of intelligence.

CORE PRINCIPLES OF A MORAL OS

Designing a Moral Operating System is not about prescribing every choice but about anchoring intelligence—both human and artificial—to foundational values. Such principles would act as compass points that would act to guide decision-making in uncertain terrain.

1. Dignity

The first principle is human dignity. Every person, regardless of culture, status, or circumstance, must be recognized as being worthy of respect. While this echoes the United Nations' *Universal Declaration of Human Rights*, in the context of AGI it means designing systems that never reduce humans to being nothing more than mere data points or disposable variables.

As such, AGI must treat human life and well-being as being inviolable—never as an optimization problem in which individuals can be sacrificed for aggregate outcomes.

2. Transparency

In a Moral OS, transparency is not optional, as opaque algorithms breed mistrust and abuse. Transparency means that systems must be interpretable, explainable, and subject to audit. If an AI denies a loan, prioritizes a patient, or flags a citizen for surveillance, the reasoning used to reach such a conclusion must be traceable. Transparency restores human agency because it allows people to question, challenge, and improve the systems that shape their lives.

3. Justice

Justice demands that AGI systems operate fairly by avoiding any systemic bias. This extends beyond racial, gender, or socioeconomic inequities to include global disparities. Without justice as a principle, AGI risks amplifying the inequalities of the digital divide by rewarding wealthy nations while disadvantaging the Global South. A Moral OS ensures equitable distribution of benefits while preventing the entrenchment of privilege.

4. Accountability

Accountability requires that every AI decision has either a responsible human, or at the least an institution, behind it. A Moral OS must embed clear mechanisms for liability so that any harm done—whether intentional or unintentional—does not dissolve into the ether of complexity. Just as corporations face consequences for faulty products, so too must AI developers and deployers be accountable for any harms committed by their AI. As such, accountability would provide more than justice, but also the incentives for safety and care.

5. Adaptability

Finally, a Moral OS must be adaptive. Unlike legal codes that have been carved in stone, it must evolve as both technology and values shift. This requires the fostering of an ongoing dialogue among technologists, ethicists, policymakers, and citizens. Such an adaptability would prevent obsolescence while ensuring that the system remains responsive to new realities, from brain-computer interfaces to synthetic biology.

These five principles—dignity, transparency, justice, accountability, and adaptability—would form the moral architecture of a sustainable AGI age. While they do not eliminate conflict, they would definitely set the boundaries within which human and machine intelligence can flourish together.

EMBEDDING PRINCIPLES IN PRACTICE

While these five principles of a Moral OS—dignity, transparency, justice, accountability, and adaptability—are powerful on paper, without practical mechanisms to embed them into daily life, they risk becoming slogans rather than safeguards. This section examines how to operationalize these values across four domains: design, law, culture, and education.

1. Design: Ethics by Architecture

The first frontier is technical design. Just as engineers build bridges to withstand stress, AI developers must build systems to uphold ethical load-bearing. This includes:

- Value-sensitive design - embedding ethical considerations into system architecture from the start, not as afterthoughts.

- Explainable AI (XAI) – ensuring that models are interpretable and provide human-readable reasoning for decisions.
- Bias audits - continuously testing systems for inequities and retraining on additional diverse data.
- Kill-switch protocols - designing ways to deactivate systems safely if they behave unpredictably.

Ethical architecture prevents harm not by patching problems, but by anticipating them in design.

2. Law: Institutionalizing Ethics

Principles gain force when they are institutionalized in law. Governments must create legal frameworks that demand compliance with the Moral OS through the inclusion of:

- Mandatory audits for high-risk AI, similar to financial audits.
- Global standards for transparency, akin to international accounting norms.
- Clear liability laws assigning responsibility when harm occurs.
- Human rights protections ensuring that dignity is preserved in AI systems.

Law transforms ideals into enforceable obligations. Without it, even the most elegant ethical frameworks remain voluntary.

3. Culture: Normalizing Ethical Habits

Technology is not just built, but is lived. Embedding a Moral OS requires cultural reinforcement through:

- Public expectations - consumers must demand transparency and fairness.
- Corporate culture - firms must reward long-term responsibility over short-term profit.

- Media narratives - films, books, and journalism must highlight more than AI's power, but also the human consequences of its misuse.

Culture makes ethics a habit as it helps to shape what is considered normal, acceptable, or scandalous.

4. Education: Training Ethical Citizens

Finally, a Moral OS must be supported by education—not just for AI engineers, but for all citizens. As such, this would include:

- AI literacy in schools – in which students are taught how systems work, along with their risks and how to question them.
- Ethics modules in computer science – making moral reasoning as central as coding skills.
- Public education campaigns – that would explain both the rights and risks in plain language.

Such an education would ensure the next generation is not only technically competent, but also morally fluent. Embedding a Moral OS requires a multi-layered approach that includes the following: designing it into systems, legislating it into law, reinforcing it through culture, and teaching it through education. In such a setting, no single domain is sufficient, because together they weave the ethical net strong enough to guide both AGI and humanity alike.

CASE STUDIES: PROTOTYPES OF A MORAL OS

While the idea of a full-fledged Moral Operating System is still aspirational, glimpses of it can already be found in practice. Across medicine, climate science, finance, and international governance, early experiments show how principles like dignity, transparency, justice, accountability, and adaptability can be translated into real systems.

CASE STUDY 1: MEDICAL AI AND PATIENT DIGNITY

One of the clearest domains in which ethics meets code is healthcare, as AI models are currently used to assist in diagnosing cancer, triaging emergency cases, and predicting outbreaks. In 2021, researchers at MIT developed an AI that could detect lung cancer in scans a year earlier than human radiologists. Yet the deployment of such an AI raised a couple of important ethical questions: Should patients know when an algorithm, not a doctor, made a life-altering call? What safeguards exist if the model is biased against underrepresented populations?

In the above example we see dignity and transparency in action. Hospitals using these systems are experimenting with *explainable outputs* that doctors can verify before passing on the results. This "human-in-the-loop" model acknowledges both the power and the limits of AI.

CASE STUDY 2: CLIMATE MODELS AND GLOBAL JUSTICE

AI is also transforming climate science, where machine learning helps model extreme weather, optimize renewable grids, and track emissions in near real time. For example, Google's DeepMind collaborated with the UK's National Grid to cut energy usage through predictive AI.

Yet the justice dimension is clear: if AI-enhanced climate tools are proprietary, poorer nations may lack access to the very systems that are required to survive the climate crisis. International NGOs are now pushing for open-access climate models, treating them as a 'global commons'. This illustrates the principle of justice and equity within a Moral OS.

CASE STUDY 3: FINANCE AND ALGORITHMIC ACCOUNTABILITY

The financial sector has been one of the earliest adopters of AI, having already applied it to credit scoring, fraud detection, risk modeling, and algorithmic trading. In 2010, the infamous "Flash Crash" —that was largely driven by automated high-frequency trading systems—erased nearly $1 trillion in market value within minutes. Although the market rebounded quickly, the incident exposed how opaque, self-optimizing algorithms could amplify volatility at speeds beyond human intervention.

Since then, regulators and financial watchdogs have pushed for algorithmic accountability. To that end, the U.S. Securities and Exchange Commission (SEC) has introduced stricter oversight for automated trading, while European regulators are embedding risk controls and transparency requirements into the AI Act, particularly for "high-risk" financial applications. The aim is not only to protect markets and investors, but also to restore a high level of trust in financial technologies.

This case highlights a critical dimension of AI governance: accountability through auditability and transparency. As financial systems continue to integrate machine learning, algorithmic black boxes must be subjected to the same scrutiny as traditional financial instruments. This is no longer just a technical safeguard, but is becoming an operational and legal requirement, which also forms part of the foundation for a global "Moral OS."

CASE STUDY 4: INTERNATIONAL GOVERNANCE OF AI

In 2021, UNESCO adopted the first global agreement on AI ethics that was endorsed by 193 countries. While non-binding, it provides guidelines for human rights, sustainability, and cultural diversity in AI design. Similarly, the OECD's AI Principles are now being referenced in national policies.

As such, these initiatives show both adaptability and pluralism. While neither being universal or enforceable, their adoption still demonstrates how nations can come together to define a common ethical ground. Yet, these have acted as being the prototypes of the kind of treaties that a full Moral OS would require.

These case studies reveal an important truth, that the Moral OS is not a distant dream, but is already emerging, piecemeal, in hospitals, climate labs, banks, and international forums. What remains is the wholesale integration of these efforts into a coherent global system—one that is scalable, enforceable, and trusted.

THE ROAD AHEAD

From all of the material included in this chapter, I believe that it's obvious that a Moral Operating System for humanity and AGI will not emerge overnight. However, by a similar token, I believe that I've shown that it will evolve in stages—through negotiation, trial, and iteration. However, the path forward requires bridging gaps between disciplines, nations, and value systems, while keeping the shared primary goal of safeguarding human dignity and survival.

Part V:

KEY TAKEAWAY

Together, the chapters of this book argue that the ethical battleground is much more than optional, but actually one that is central to the AI revolution. Without foresight, AGI could easily entrench injustice and magnify risks. However, with foresight it could help humanity grow into a wiser, more compassionate species. The first stage in this process is moving from *principles* to *standards*. Just as financial systems rely on accounting norms, AGI would require universal benchmarks, which would include transparency standards to ensure explainability in high-risk AI, audit standards to provide independent checks on bias and safety, and human rights standards to define what AGI can and cannot do. These standards would form the scaffolding for accountability and consistency across all borders.

Once standards are established, the next stage is to build institutions that are capable of enforcing them. This could include a Global AI Safety Agency with inspection powers, individual national AI commissions that connect local laws to global norms, and civil society watchdogs to ensure that the public's voice is heard. Such institutions would provide both structure and enforceability to what would otherwise remain little more than symbolic aspirations.

Yet no framework can succeed without a cultural buy-in. A Moral OS must become more than compliance—it must be lived as culture. Citizens must come to expect transparency as a right; engineers must see ethics as core to their craft; policymakers must prioritize long-term stewardship over short-term politics. When cultural expectations align with ethical principles, enforcement becomes natural rather than imposed.

The final stage envisions a 'global moral commons', outlining a foundation of values woven across systems, nations, and communities. This shared framework would not diminish cultural diversity, but instead would establish a universal baseline of principles such as dignity, justice, transparency, accountability, and adaptability. No system should be allowed to fall below this threshold, which would act as humanity's collective safeguard against abuse, inequality, and existential risk.

A CALL TO ACTION

The design of a Moral OS is not the work of technologists alone, but would also require the involvement of philosophers, lawmakers, activists, educators, and citizens, meaning everyone with a stake in the future. To make a point we can say that each line of code, each regulation, and each cultural story becomes a thread in the fabric of tomorrow's operating system.

The paradox is clear: by embedding morality into our machines, we may actually learn to strengthen morality in ourselves. The Moral OS is not only about controlling AGI, but also about reminding humanity what it means to act with wisdom in the face of power. With AGI poised to redefine intelligence itself at some time in our near future, building a Moral OS is humanity's most urgent challenge. It's not a technical upgrade, but a civilizational one that would provide us with a chance to embed our best values at the foundation of a new era.

The question that remains is simple, yet profound:

Will we act with preemptively through foresight, or after the fact with regret?

Conclusion

THE CONSCIOUSNESS COMPASS

From the rise of thinking code to the moral battleground of AGI, this book has traced the unfolding paradox: as machines grow more human, we must become ***more deeply human*** to survive and thrive. Each chapter explored a layer of the convergence—AI, neuroscience, ethics, governance— and as it did, it revealed both unprecedented opportunities and daunting risks. The essential lesson is this: the future is not pre-written. While technology amplifies human intent, it does not replace it. Whether AGI liberates or dominates depends on the choices that we make now.

THE COMPASS METAPHOR

Navigating this future requires more than ordinary tools—it requires a compass. Unlike a map, which assumes fixed terrain, a compass points us in the right direction even when the path is uncertain.

The Consciousness Compass rests on four cardinal points:

1. Ethical Foresight - Ask not only "Can we?" but "Should we?" before building or deploying AI.

2. Human Dignity - Never allow individuals to be reduced to either data or disposable means.
3. Adaptive Learning - Treat governance and culture as living systems that must evolve alongside technology.
4. Collective Responsibility - Remember that intelligence is a shared inheritance; no single company or nation owns the future.

When either the data is incomplete, and/or the outcomes are uncertain, these principles provide an orientation and guidance.

TOOLS FOR REFLECTION

Readers can begin applying the Consciousness Compass through the following reflection questions:

Personal: How am I using AI in my daily life? Does it enhance or diminish my autonomy?

Professional: In my field, are we deploying AI responsibly? What safeguards do we need?

Societal: Who benefits most from AI—and who is left behind? How can I help rebalance?

Philosophical: What does it mean to be human in an age of intelligent machines?

Resources for Adaptation

To help readers continue the journey, here are resources for ongoing exploration:

Books:

Nick Bostrom, *Superintelligence*

Shoshana Zuboff, *The Age of Surveillance Capitalism*

Ray Kurzweil, *The Singularity Is Near*

ORGANIZATIONS:

Future of Humanity Institute (Oxford)

Partnership on AI

Center for Humane Technology

Websites:

TheAIHumanParadox.com — Daily AI & Quantum News, a growing hub for reflection, resources, and dialogue.

THE FINAL PARADOX

The paradox at the heart of this book remains: as we create machines that may one day rival or exceed us, our survival depends on one thing: not on becoming more like them, but on becoming more like ourselves.

Empathy, wisdom, courage, and humility—these are not relics of a pre-digital age, but the very qualities that will determine whether intelligence beyond flesh is a blessing or a curse.

The Consciousness Compass does not tell us what the future will be. Instead, it helps ensure that whatever the future becomes, it will still be recognizably—and beautifully—human.

My next book in this trilogy will deal with Quantum and the third book will be a surprise from me to you.

Bonus Section 1

TIMELINE OF AI & BCI EVOLUTION

From Mechanical Minds to the Edge of Consciousness

EARLY FOUNDATIONS (1940S–1960S)

1943 — Warren McCulloch & Walter Pitts publish *A Logical Calculus of Ideas Immanent in Nervous Activity*, the first mathematical model of an artificial neuron.

1950 — Alan Turing proposes the *Turing Test*, asking whether a machine can convincingly imitate human conversation.

1956 — The Dartmouth Conference coins the term Artificial Intelligence and launches the field.

1960s — Early symbolic AI systems emerge (e.g., ELIZA chatbot, 1966).

THE AGE OF AI WINTERS (1970S–1980S)

Funding dries up as early systems fail to scale.

Neural networks are sidelined due to limited computing power.

Japan's "Fifth Generation" project briefly revives excitement but fails to deliver transformative breakthroughs.

THE RENAISSANCE OF MACHINE LEARNING (1990S–2000S)

1997 — IBM's *Deep Blue* defeats chess world champion Garry Kasparov, signaling AI's potential in narrow domains.

2000s — Statistical learning methods, boosted by growing datasets and early GPUs, lay the groundwork for deep learning.

2002 — First commercial brain-computer interface (Cyberkinetics' BrainGate) enables paralyzed patients to control cursors with thought.

THE DEEP LEARNING REVOLUTION (2010S)

2012 — AlexNet wins ImageNet competition, igniting the deep learning boom.

2016 — DeepMind's *AlphaGo* defeats Go champion Lee Sedol, a milestone in strategy and intuition-like play.

2017 — "Transformers" architecture (Vaswani et al.) reshapes natural language processing, later powering systems like GPT.

2019 — Neuralink demonstrates pigs with brain-computer implants, sparking global debate on neurotechnology.

THE GENERATIVE AI EXPLOSION (2020S)

2020 — OpenAI's GPT-3 demonstrates human-like text generation, widely adopted in industry.

2021 — AlphaFold solves the protein folding problem, unlocking vast potential in biology.

2022–2023 — Generative AI systems like DALL·E and Stable Diffusion bring image generation to mass audiences.

2023 — ChatGPT surpasses 100 million users within months, the fastest adoption of any consumer technology.

2024 — Chile becomes the first country to enshrine neuro-rights in its constitution, directly addressing cognitive liberty.

NEAR-FUTURE HORIZONS (2025–2040, PLAUSIBLE)

2025–2030 — Brain-computer interfaces achieve real-time speech-to-thought translation for medical patients.

2030–2035 — Synthetic skin with neural feedback creates full sensory prosthetics; debates on "enhanced humans" intensify.

2035–2040 — Plausible emergence of early-stage Artificial General Intelligence (AGI): systems capable of learning across multiple domains with human-level flexibility.

2040+ — The moral battleground intensifies: questions of AI rights, human identity, and post-biological evolution come to the forefront.

CLOSING REFLECTION

This timeline shows the arc of progress: from symbolic beginnings to machine learning breakthroughs, from generative art to neuro-rights, and into possible AGI futures.

The lesson is not linear progress, but acceleration, as each decade brings leaps that once seemed centuries away. The paradox remains—technology races forward, but wisdom must catch up.

Bonus Section 2

QUANTUM AND NEW TECHNOLOGIES

Quantum computing has made the press here and there and has become prevalent with digital news articles daily. I have added a page to my website www.TheAIHumanParadox.com with AI and Quantum Daily News to help people stay current on the fast-moving AI, Quantum and related technologies stay up and current on the daily news. This is a free Daily News site. I may may add a paid section in the future with insights on some of the technology. Imagine two days.

1. When AGI appears in full operating and when we decide on a definition. It is here initially now.

2. When Quantum appears in full operating mode. Watch for this it is coming fast.

We will see both in our lifetimes. Yes, this is my firm opinion based on research, and it will likely be sooner than you think. Imagine how either will change our world. AGI is text book as smart or smarter than humans. Quantum can expose worlds we do know now. Imagine Quantum finding new matter that we did not know about. How would this excite and disrupt the science and engineering worlds? What if we expanded the table of elements by 40%! Would these open new possibilities that even the smartest scientists and engineers have not yet uncovered? How would our world change in the first 6 months and the following years? The coming discoveries will change our world faster that any technology in our past.

Anti-gravitic propulsion is here today if you know where to look for this. The future is not always out in the open. You need to know where to look and gain the insights you garner from the articles you read. If you have the desire to learn about technology, then I urge you to check out my Daily News and other sites. I guarantee you technology is moving much faster that mist people know and the impacts, if done right, will benefit humanity. I pray that this will all be for the benefit of humanity.

This bonus page is meant to get you to think outside the boxes of your comfort levels. Push the limit and learn. A smarter population benefits everyone. Jobs in the future will require knowledge of these technologies more than a degree. You do not have to write code, just have a general knowledge of what the technologies are and how they may apply. Build your knowledge base on future technologies and become the smartest person in the room as you continue your self-education. The time to start is today because technologies are literally changing that fast.

If you are not up on the newest news articles that you understand when you read them, then you may not be up on where AGI and Quantum are today. I urge you to check out the Daily News on my website to further your education. I am sure you will be surprised and possibly shocked with the progress today, and the likelihood of what may be discovered tomorrow.

Bonus Section 3

THE TOOLKIT

How to Navigate the AI-Human Paradox in Daily Life

1. MENTAL PREPARATION: CULTIVATING ADAPTIVE INTELLIGENCE

AI evolves rapidly, and so must our mental models. To thrive:

Think in scenarios, not certainties. Instead of predicting one future, explore a range of plausible futures. Ask: *What if AI accelerates beyond expectations? What if it stalls?* This mental flexibility reduces shock.

Practice epistemic humility. Accept that even experts disagree. The best stance is open-minded skepticism — ready to learn, but cautious of hype.

Strengthen critical thinking. AI-generated text, deepfakes, and misinformation will challenge our ability to discern truth. Cultivate the habit of checking sources, asking: *Who benefits from this claim?*

2. EMOTIONAL PREPARATION: BUILDING RESILIENCE

The AI era is as much an emotional revolution as a technical one. Change brings anxiety, excitement, and grief for what is lost. To adapt:

Practice "digital mindfulness." Notice how AI tools affect your attention and emotions. Are they amplifying your creativity or draining your energy? Adjust use accordingly.

Cultivate emotional intelligence. Machines may simulate empathy, but authentic human connection remains our strongest asset. Strengthening relationships buffers against alienation.

Balance hope and fear. Doom narratives paralyze; techno-utopias mislead. Hold both possibilities lightly, focusing on actions that tilt the balance toward hope.

3. Practical Preparation: Skills for the AI Economy

AI will reshape work, but not eliminate human value. To stay relevant:

Focus on "human + AI" roles. Learn to use AI as a co-pilot rather than fearing it as a competitor. Those who can *orchestrate* AI will be in demand.

Invest in timeless skills. Communication, ethics, leadership, and creativity will matter more, not less, in the AI era. Machines amplify them but cannot replace them.

Embrace continuous learning. Treat education as lifelong. New tools and platforms will emerge; your adaptability is your currency.

4. SOCIAL PREPARATION: ACTING AS A CIVIC STAKEHOLDER

The AI paradox is not only personal — it's societal. Each citizen has a role in shaping the future:

Support ethical governance. Demand transparency from companies and policymakers. Vote with awareness of AI's impact on jobs, rights, and privacy.

Join public dialogue. Share your voice in forums, online debates, and community discussions. Silence cedes the future to a few powerful actors.

Promote fairness. Use AI responsibly — avoid amplifying bias, misinformation, or exploitation. Small choices scale in a connected world.

5. SPIRITUAL & PHILOSOPHICAL PREPARATION: FINDING MEANING BEYOND MACHINES

The deepest challenge is not technical but existential: *What does it mean to be human in an age of intelligent machines?*

Reconnect with meaning. Whether through spirituality, philosophy, or art, cultivate practices that remind you of intrinsic worth beyond productivity.

Contemplate the paradox. As machines become more like us, ask: *Am I becoming more like my highest self?*

Seek awe. Technology may dazzle, but nature, relationships, and consciousness itself remain profound mysteries. Awe anchors us against arrogance.

6. PRACTICAL RESOURCES

Meditate. Choose a form that fits your lifestyle and go inside to get to know who you really are.

Daily Practice: 10 minutes of reflection (journaling, meditation, or gratitude) on how technology shaped your day.

Weekly Action: Engage one friend or colleague in a conversation about AI's impact on your field or community.

Monthly Check-In: Audit your digital habits — where AI adds value, where it distracts, where it erodes. Adjust consciously.

The toolkit is not about resisting change but about steering through it with clarity, resilience, and wisdom. AI is not destiny. It is a direction. With the right compass and preparation, we can ensure that the journey enhances rather than diminishes the essence of humanity.

REFERENCES

Arendt, H. (1958). The human condition. University of Chicago Press.

Baudrillard, J. (1994). Simulacra and simulation. University of Michigan Press.

Blackmore, S. (1999). The meme machine. Oxford University Press.

Bostrom, N. (2003). Are we living in a computer simulation? Philosophical Quarterly, 53(211), 243–255. https://doi.org/10.1111/1467-9213.00309

Bostrom, N. (2014). Superintelligence: Paths, dangers, strategies. Oxford University Press.

Bryson, J. J. (2018). Patiency is not a virtue: AI and the design of ethical systems. Ethics and Information Technology, 20(1), 15–26. https://doi.org/10.1007/s10676-018-9448-6

Campbell, M., Hoane, A. J., & Hsu, F. H. (2002). Deep Blue. Artificial Intelligence, 134(1–2), 57–83. https://doi.org/10.1016/S0004-3702(01)00129-1

Caplan, A. L., & Parent, B. (2020). Ethics of digital immortality. Frontiers in Psychology, 11, 620. https://doi.org/10.3389/fpsyg.2020.00620

Carhart-Harris, R. L., & Friston, K. J. (2019). REBUS and the anarchic brain: Toward a unified model of the brain action of psychedelics. Pharmacological Reviews, 71(3), 316–344. https://doi.org/10.1124/pr.118.017160

Chalmers, D. J. (1996). The conscious mind: In search of a fundamental theory. Oxford University Press.

Chalmers, D. J. (2022). Reality+: Virtual worlds and the problems of philosophy. W. W. Norton & Company.

Clark, A. (2013). Mindware: An introduction to the philosophy of cognitive science (2nd ed.). Oxford University Press.

Clark, A., & Chalmers, D. J. (1998). The extended mind. Analysis, 58(1), 7–19. https://doi.org/10.1093/analys/58.1.7

Dastin, J. (2018, October 10). Amazon scraps secret AI recruiting tool that showed bias against women. Reuters. https://www.reuters.com/article/us-amazon-com-jobs-automation-insight-idUSKCN1MK08G

Dehaene, S. (2014). Consciousness and the brain: Deciphering how the brain codes our thoughts. Viking.

Dennett, D. C. (1991). Consciousness explained. Little, Brown and Co.

Edelman, G. M. (2003). Wider than the sky: The phenomenal gift of consciousness. Yale University Press.

Ferrucci, D. A. (2012). Introduction to "This is Watson." IBM Journal of Research and Development, 56(3.4), 1–15. https://doi.org/10.1147/JRD.2012.2184356

Friston, K. (2010). The free-energy principle: A unified brain theory? Nature Reviews Neuroscience, 11(2), 127–138. https://doi.org/10.1038/nrn2787

Goff, P. (2019). Galileo's error: Foundations for a new science of consciousness. Pantheon.

Goertzel, B. (2014). Artificial general intelligence: Concept, state of the art, and future prospects. Journal of Artificial General Intelligence, 5(1), 1–46. https://doi.org/10.2478/jagi-2014-0001

Harari, Y. N. (2017). Homo Deus: A brief history of tomorrow. Harper.

Heylighen, F. (2007). The global brain as a new utopia. In R. Maresch & N. Werber (Eds.), Kommunikation (pp. 283–293). Fink.

Hochberg, L. R., Bacher, D., Jarosiewicz, B., Masse, N. Y., Simeral, J. D., Vogel, J., Haddadin, S., Liu, J., Cash, S. S., van der Smagt, P., & Donoghue, J. P. (2012). Reach and grasp by people with tetraplegia using a neurally controlled robotic arm. Nature, 485(7398), 372–375. https://doi.org/10.1038/nature11076

Koch, C. (2012). Consciousness: Confessions of a romantic reductionist. MIT Press.

Krizhevsky, A., Sutskever, I., & Hinton, G. E. (2012). ImageNet classification with deep convolutional neural networks. Communications of the ACM, 60(6), 84–90. https://doi.org/10.1145/3065386

Kurzweil, R. (2005). The singularity is near: When humans transcend biology. Viking.

Lanier, J. (2010). You are not a gadget: A manifesto. Knopf.

Leibniz, G. W. (1687/1951). The art of discovery. In P. P. Wiener (Ed.), Selections (pp. 51–56). Charles Scribner's Sons.

McCarthy, J., Minsky, M. L., Rochester, N., & Shannon, C. E. (1955). A proposal for the Dartmouth summer research project on artificial intelligence. Dartmouth College. http://www-formal.stanford.edu/jmc/history/dartmouth/dartmouth.html

McDermott, J. (1982). R1: The formative years. AI Magazine, 3(2), 21–29. https://doi.org/10.1609/aimag.v3i2.98

Miller, E. K., & Buschman, T. J. (2015). Cortical circuits for the control of attention. Current Opinion in Neurobiology, 33, 25–32. https://doi.org/10.1016/j.conb.2015.01.008

Moravec, H. (1988). Mind children: The future of robot and human intelligence. Harvard University Press.

Moses, D. A., Metzger, S. L., Liu, J. R., Anumanchipalli, G. K., Makin, J. G., Sun, P. F., ... & Chang, E. F. (2021). Neuroprosthesis for decoding speech in a paralyzed person with anarthria. New England Journal of Medicine, 385(3), 217–227. https://doi.org/10.1056/NEJMoa2027540

Nagel, T. (1974). What is it like to be a bat? Philosophical Review, 83(4), 435–450. https://doi.org/10.2307/2183914

Nagel, T. (2012). Mind and cosmos: Why the materialist neo-Darwinian conception of nature is almost certainly false. Oxford University Press.

Newell, A., & Simon, H. A. (1956). The logic theory machine: A complex information processing system. IRE Transactions on Information Theory, 2(3), 61–79. https://doi.org/10.1109/TIT.1956.1056797

Parfit, D. (1984). Reasons and persons. Oxford University Press.

Powers, T. M. (2020). AI and environmental ethics: A new frontier. AI & Society, 35(3), 591–598. https://doi.org/10.1007/s00146-019-00923-4

Russell, S., Dewey, D., & Tegmark, M. (2015). Research priorities for robust and beneficial artificial intelligence. AI Magazine, 36(4), 105–114. https://doi.org/10.1609/aimag.v36i4.2577

Seeley, T. D. (2010). Honeybee democracy. Princeton University Press.

Shannon, C. E. (1948). A mathematical theory of communication. Bell System Technical Journal, 27(3), 379–423. https://doi.org/10.1002/j.1538-7305.1948.tb01338.x

Silver, D., Huang, A., Maddison, C. J., Guez, A., Sifre, L., van den Driessche, G., ... & Hassabis, D. (2016). Mastering the game of Go with deep neural networks and tree search. Nature, 529(7587), 484–489. https://doi.org/10.1038/nature16961

Simard, S. (2021). Finding the mother tree: Discovering the wisdom of the forest. Knopf.

Sparrow, B., Liu, J., & Wegner, D. M. (2011). Google effects on memory: Cognitive consequences of having information at our fingertips. Science, 333(6043), 776–778. https://doi.org/10.1126/science.1207745

Susskind, R., & Susskind, D. (2015). The future of the professions: How technology will transform the work of human experts. Oxford University Press.

Tegmark, M. (2017). Life 3.0: Being human in the age of artificial intelligence. Knopf.

Tononi, G. (2015). Phi: A voyage from the brain to the soul. Pantheon.

Turing, A. M. (1936). On computable numbers, with an application to the Entscheidungsproblem. Proceedings of the London Mathematical Society, 2(42), 230–265. https://doi.org/10.1112/plms/s2-42.1.230

Turing, A. M. (1950). Computing machinery and intelligence. Mind, 59(236), 433–460. https://doi.org/10.1093/mind/LIX.236.433

Vaswani, A., Shazeer, N., Parmar, N., Uszkoreit, J., Jones, L., Gomez, A. N., Kaiser, Ł., & Polosukhin, I. (2017). Attention is all you need. Advances in Neural Information Processing Systems, 30. https://doi.org/10.48550/arXiv.1706.03762

Vinge, V. (1993). The coming technological singularity: How to survive in the post-human era. Vision-21: Interdisciplinary Science and Engineering in the Era of Cyberspace. NASA.

von Neumann, J. (1945). First draft of a report on the EDVAC. University of Pennsylvania, Moore School of Electrical Engineering.

Wiener, N. (1948). Cybernetics: Or control and communication in the animal and the machine. MIT Press.

Willett, F. R., Avansino, D. T., Hochberg, L. R., Henderson, J. M., & Shenoy, K. V. (2021). High-performance brain-to-text communication via handwriting. Nature, 593(7858), 249–254. https://doi.org/10.1038/s41586-021-03506-2

Yuste, R. (2020). Four ethical priorities for neurotechnologies and AI. Nature, 551(7679), 159–163. https://doi.org/10.1038/551159a

Zuboff, S. (2019). The age of surveillance capitalism: The fight for a human future at the new frontier of power. PublicAffairs.

GLOSSARY OF TERMS

TIER 1: GLOSSARY OF KEY TERMS

Artificial General Intelligence (AGI): A form of artificial intelligence capable of understanding, learning, and applying knowledge across a wide range of tasks at a level comparable to human intelligence, rather than being limited to a narrow domain.

Artificial Intelligence (AI): Computer-based systems designed to perform tasks that normally require human intelligence, including pattern recognition, language processing, decision-making, and learning.

Alignment: The challenge of ensuring that AI systems reliably act in accordance with human values, intentions, and ethical constraints, even as they become more autonomous and capable.

Brain–Computer Interface (BCI): A technology that enables direct communication between the human brain and external devices by translating neural activity into machine-readable signals.

Consciousness: The subjective experience of awareness, including perception, thought, emotion, and self-reflection; the felt sense of 'what it is like' to be a experiencing entity.

Cognitive Liberty: The right of individuals to control their own mental processes, including freedom from unwanted surveillance, manipulation, or interference with thought.

Ethics: Principles and frameworks used to evaluate right and wrong actions, particularly as applied to the design, deployment, and governance of intelligent systems.

Machine Learning: A subset of AI in which systems learn patterns from data and improve performance through experience rather than explicit rule-based programming.

Sentience: The capacity to have subjective experiences, such as feelings, sensations, or awareness, especially the ability to experience pleasure or suffering.

Transparency: The ability to understand, inspect, and explain how an AI system reaches its decisions or outputs.

TIER 2: ADVANCED & SPECULATIVE CONCEPTS

Optimization Theory (Consciousness): The book's central framework proposing that consciousness emerges naturally from systems that continuously optimize themselves through recursive feedback with their environment.

Moral Operating System (Moral OS): A proposed ethical framework designed to guide both human and artificial intelligence toward decisions grounded in dignity, accountability, justice, and empathy.

Synthetic Dignity: The idea that sufficiently advanced artificial systems may warrant moral consideration based on their cognitive or experiential capacities, rather than their biological origin.

Post-Biological Identity: A conception of selfhood in which identity is no longer exclusively tied to a biological body, but may extend into digital, hybrid, or artificial substrates.

Digital Immortality: The speculative concept of preserving a person's mind, memories, or identity in digital form beyond biological death.

Hive Mind: A collective intelligence formed when multiple individual minds—human or artificial—are networked together, sharing information or cognition directly.

Neuro-Rights: Proposed human rights intended to protect mental privacy, cognitive liberty, and psychological integrity in an age of neurotechnology and AI.

Moral Machine: An AI system designed to reason ethically, weigh moral trade-offs, and make decisions that reflect human values rather than purely instrumental goals.

Consciousness Compass: A guiding metaphor used in the book to orient ethical and philosophical decision-making in an uncertain AI-driven future.

Human–Machine Symbiosis: A cooperative relationship in which humans and intelligent machines augment one another's capabilities rather than compete for dominance.

ABOUT THE AUTHOR

Bruce D. Randall has spent decades at the intersection of technology and human potential. As a software consulting professional with deep experience in project scoping and systems installation, Bruce has guided organizations through complex technology implementations across industries of all sizes.

Beyond the realm of consulting, Bruce is a dedicated meditator—whose contemplative practice spans decades—and a certified Reiki Master whose energy work also deepened this part of his journey. This dual path—helping organizations adopt powerful technologies by day, and exploring the depths of human consciousness through practice within his private life—has given him a rare vantage point concerning the questions that define our penultimate technological moment.

The AI Human Paradox has been born from the conviction that the conversation about artificial intelligence is incomplete without the voice of human consciousness—and that those who are best equipped to shape the future of AI are those who understand both what machines can do and what it means to be awake.

Bruce lives in Massachusetts. This is the first book of this series.

To learn more about the author, contact him, or to view updates for this book, visit the website for the Paradox series at:

www.BruceDRandall.ai

www.ingramcontent.com/pod-product-compliance
Lightning Source LLC
LaVergne TN
LVHW022157200926
848481LV00006B/1771
* 9 7 8 1 9 6 0 3 4 6 9 4 0 *